Theoretical Biology and Complexity

Three Essays on
the Natural Philosophy
of Complex Systems

Contributors

A. H. Louie
I. W. Richardson
Robert Rosen

Theoretical Biology and Complexity

Three Essays on the Natural Philosophy of Complex Systems

Edited by

Robert Rosen

Department of Physiology and Biophysics
Dalhousie University
Halifax, Nova Scotia

1985

ACADEMIC PRESS, INC.

(Harcourt Brace Jovanovich, Publishers)

Orlando San Diego New York London
Toronto Montreal Sydney Tokyo

ACADEMIC PRESS, INC.
Orlando, Florida 32887

United Kingdom Edition published by
ACADEMIC PRESS INC. (LONDON) LTD.
24–28 Oval Road, London NW1 7DX

Library of Congress Cataloging in Publication Data
Main entry under title:

Theoretical biology and complexity.

Contents: The dynamics and energetics of complex real systems / I. W. Richardson -- Categorical system theory / Aloisius Ho-Yin Louie -- Organisms as causal systems which are not mechanisms / Robert Rosen.
1. Biological systems--Addresses, essays, lectures. 2. Biological systems--Measurement--Addresses, essays, lectures. 3. System theory--Addresses, essays, lectures. I. Richardson, I. W., DATE. Dynamics and energetics of real systems. 1985. II. Louie, Aloisius Ho-Yin. Categorical system theory. 1985. III. Rosen, Robert. Organisms as causal systems which are not mechanisms. 1985.
QH311.T48 1985 574 84-21568
ISBN 0–12–597280–6 (alk. paper)

PRINTED IN THE UNITED STATES OF AMERICA

85 86 87 88 9 8 7 6 5 4 3 2 1

Contents

3 Organisms as Causal Systems Which Are Not Mechanisms: An Essay into the Nature of Complexity

Robert Rosen

List of Contributors

Numbers in parentheses indicate the pages on which the authors' contributions begin.

*A. H. Louie** (69), Department of Physiology and Biophysics, Dalhousie University, Halifax, Nova Scotia B3H 4H7, Canada

I. W. Richardson† (1), Department of Physiology and Biophysics, Dalhousie University, Halifax, Nova Scotia B3H 4H7, Canada

Robert Rosen (165), Department of Physiology and Biophysics, Dalhousie University, Halifax, Nova Scotia B3H 4H7, Canada

* Present address: 86 Dagmar Avenue, Vanier, Ontario K1L 5T4, Canada. No longer affiliated with Dalhousie University.

† Present address: 908 Middle Avenue, Menlo Park, California 94025. No longer affiliated with Dalhousie University.

Preface

This volume is made up of three short essays, each separately conceived and written, each with distinct thrusts and emphases, but nevertheless closely related in substance and spirit. The purpose of this brief introduction is to describe some of these relations, which are both personal and scientific.

Let us turn first to the scientific threads that relate the three contributions. Although very different in emphasis and in thrust, they all spring from a common concern: to grasp and comprehend the material basis of living systems. I believe that we each began with a conviction that contemporary physics already contained the necessary universals with which to cope with the phenomena of life and that therefore only a clever rearrangement and redeployment of these universals would suffice to bring them to bear effectively on biology. I believe that we each came separately, and with great reluctance, to admit the possibility that this conviction might not be true, and hence that a true theory of thc organism required new physics and new epistemology. And again separately, we realized that the measurement process, which lies at the very heart of every mode of system description, provides perhaps the only safe and fundamental point of departure for building a comprehensive theory, not only of organisms, but of natural systems in general. This premise is the primary thread that binds the essays in this volume together. As Dr. Richardson likes to say, such an approach restores to our fragmented sciences the kind of integration and unity they possessed in an earlier time, when scientists regarded themselves as natural philosophers.

In his essay, Dr. Richardson casts his analysis of the measurement process into an elegant dualism relating modes of description and explores the consequences of this really remarkable dualism for what we may call classical physics. But the dualism he develops pertains to any kind of description; rich as it is in the familiar contexts of classical physics, it becomes still richer when treated as a universal principle and brought to bear on organisms.

Dr. Louie explores the deeper consequences of representing the properties of natural systems through states built up out of observable quantities and the dynamics that such systems impose on each other through interactions. The natural mathematical universe for exploring these consequences is category theory, a mathematical discipline that is in a formal sense the general theory of modelling and that in this application provides both the mathematical tool and the best example for exploring relationships between systems. His work provides the natural bridge between Dr. Richardson's essay and my own; he explicitly describes a number of these relationships in the course of his development.

In my own essay, a rather radical viewpoint is adopted, motivated from several different considerations. One of these is my continuing involvement with relational biology, which is described in some detail. Another was my concern with systems that can anticipate, more specifically, ones that contain internal predictive models of themselves and/or their environments, whose predictions can be utilized to modify and control present activities. Still another is an attempt to construct a dictionary relating the language of physics (forces, potentials, fields) with the informational language (code, program, computation) so characteristic of biology, by using the concept of stability as Rosetta Stone. Consideration of these problems, separately and together, led ultimately to the suggestion that our traditional modes of system representation, involving fixed sets of states together with imposed dynamical laws, strictly pertains only to an extremely limited class of systems (which I call simple systems or mechanisms). Systems not in this class I call complex, and these can only be in some sense approximated, locally and temporally, by simple ones. Such a radical alteration of viewpoint leads to a large number of concrete, practical consequences, some of which are described in the essay.

All of these essays are the products of their authors' association with the Biomathematics Program of the Department of Physiology and Biophysics at Dalhousie University in Halifax, Nova Scotia. This program has become known as the Red House because it is housed in quarters that are, in fact, red. The creation of this program was largely the work of one of the authors, I. W. Richardson. Through his initiative, effort, and energy, he was able almost single-handedly to carve out, for a perhaps brief but precious time, what I regard as one of the most innovative and fruitful programs for research and teaching in theoretical biology in the world.

I became associated with the Red House in 1975 when, again largely through Dr. Richardson's efforts, I was offered a Killam Professorship at Dalhousie University. I found that Dr. Richardson had created an ideal atmosphere, an ambience that almost compelled those who experienced it to think, study, learn, and write beyond the edge of the known. I had experienced such ambiences before, at Chicago under Rashevsky and at Buffalo under J. F. Danielli. There are others, but they are few.

The Red House program has over the course of time attracted some extraordinary students. One of these is Dr. Aloisius Louie, whose contribution to this volume is essentially his doctoral dissertation. I would normally strongly discourage any student, however bright, from undertaking a dissertation involving a major epistemological component; dissertations are risky enough without that. But Dr. Louie is a very special case; he thrives on such risks and languishes without them. Indeed, it will be seen from the definitive discussion he presents of the category-theoretic roots of system analysis that he has been from the outset a colleague rather than a student.

It is thus with a certain sense of pride that we collectively offer this volume, not only for what novel scientific material it contains, but also as a specimen of the output of the Red House program and as a symbol for what such programs can accomplish. This last is important, for it is precisely the scientific strengths of such programs that also make them vulnerable, in constant threat of engulfment by the sands of the vast academic deserts which surround them.

We hope that these essays will provide the reader with some food for thought and will convey some of the electricity and excitement, as well as the practical import, of doing theory.

1

The Dynamics and Energetics of Complex Real Systems

I. W. RICHARDSON[†]
Department of Physiology and Biophysics
Dalhousie University
Halifax, Nova Scotia

[†] Present address: 908 Middle Avenue, Menlo Park, California. No longer affiliated with Dalhousie University.

ISBN 0-12-597280-6

I. INTRODUCTION

This essay is about the interaction between particles and the consequent observable manifestations. The word *observable* carries with it two implications. From the empirical viewpoint, it implies a scale, a given level of resolution of the measuring instruments that, by coupling energetically to the system, provide information about the state of the system. From the theoretical viewpoint, it implies a complementary resolution of energy into discrete classes, in particular into energies directly available to do work, energies in bonds, and thermal energy.

Framing a physical description of a system of particles and their interactions must start with a consideration of the hierarchical structure of possible descriptions. A possible description exists for each measuring instrument that might be applied to the system (or each class of instruments having in common a given resolution). The scale made by ordering the observational resolutions of all such possible instruments (or classes of instruments) establishes a hierarchy. By means of an example, Feynman *et al.* (1963) make it clear that basic concepts such as potential energy, kinetic energy, conservative force, and identification of a particle assume precision only in the context of such a hierarchy:

> We have spent a considerable time discussing conservative forces; what about nonconservative forces? We shall take a deeper view of this than is usual, and state that there are no nonconservative forces! As a matter of fact, all the fundamental forces in nature appear to be conservative. This is not a consequence of Newton's laws. In fact, so far as Newton himself knew, the forces could be nonconservative, as friction apparently is. When we say friction *apparently is*, we are taking a modern view, in which it has been discovered that all the deep forces, the forces between the particles at the most fundamental level, are conservative.
>
> If, for example, we analyze a system like the great globular star cluster that we saw a picture of, with the thousands of stars all interacting, then the formula for the total potential energy is simply one term plus another term, etc., summed over all pairs of stars, and the kinetic energy is the sum of the kinetic energies of all the individual stars. But the globular cluster as a whole is drifting in space too, and, if we were far enough away from it and did not see the details, could be thought of as a single object. Then if forces were applied to it, some of those forces might end up driving it forward as a whole, and we would see the center of the whole thing moving. On the other hand, some of the forces can be, so to speak, "wasted" in increasing the kinetic or potential energy of the "particles" inside. Let us suppose, for instance, that the action of these forces expands the whole cluster and makes the particles move faster. The total energy of the whole thing is really conserved, but seen from the outside with our crude eyes which cannot see the confusion of motions inside, and just thinking of the kinetic energy of the motion of the whole object as though it were a single particle, it would appear that energy is not conserved, but this is due to a lack of appreciation of what it is that we see. [From Feynman *et al.* (1963), Vol. I, p. H-6.]

The specification of what entity will be deemed a particle is subjective since it is the theoretician who decides at what level of resolution the system dynamics will be formulated. For the celestial mechanician, stars and planets are particles, whereas for the cosmologist, a galaxy is a particle in an expanding gas. If one chooses a coarse level of resolution that cannot discern the motions of the constituent stars in a galaxy, then the energetics of analytical mechanics must be augmented. There must not be any inexplicable losses of energy as the galaxy moves under the influence of imposed forces. New state variables must be introduced so that a complete accounting of energy is possible. This augmentation has much in common with that required to give a proper energy accounting for a block of wood sliding down a ramp. Some of its potential and kinetic energy is degraded by friction to motions that cannot be resolved by instruments that see the block as particle. In short, the existence of a hierarchy of descriptions necessitates the science of thermodynamics.

The words *particle* and *interaction* conjure up a duality, one which is very real and must not be ignored. The mathematical representation of interaction is achieved by fields, and since Maxwell's time physicists have considered fields to be as substantial as particles. In fact, some theories have reduced the primacy of the substantial particle to nothing more than a knot in the field. For our purposes, however, it is best not to go to such an extreme. Field and particle will have equal status. Furthermore, neither will assume meaning outside the context of the other.

Consider, for example, a particle in the earth's gravitational field. Within the limits imposed by having to make all observations of the system by means of a beam balance, two particles are identical if they respond to the field in the same manner. It is said that they have equal mass. Particle identity is established by comparison of behavior in a given field. On the other hand, two fields are identical if a given particle exhibits identical behavior in each of the fields. In a conjugate way, mass measures field and field measures mass. The duality becomes complete upon recognizing that the field has no *a priori* existence but is generated by the disposition of mass in space. Likewise, mass has no *a priori* existence since it is operationally impossible to assign to a particle a mass value in the absence of a gravitational field, that is, in the absence of an interaction with another gravitating body.

Mathematically there needs be a congruence between the world described as particle and the world described as interaction. There are scale factors that provide a bridge, a mapping, between variables of these two modes of description. These mappings and the congruences between descriptions that must necessarily be obeyed are central to the following analysis of complex real systems. The systems are of sufficient complexity to require the use of field concepts and the mathematics of the continuum. They are real, in contrast to ideal, in that dissipative processes are allowed—or rather, must be allowed since the viewpoint is hierarchical and the description is deemed to be complete under all modalities of interaction.

The scale factors and the mappings they effect make for a formal, systematic presentation. However, it must not be assumed that this is an axiomatic treatment. It is more like a sketch, with the details left for the reader to fill in. The reader must have an acquaintance with elementary classical physics and a reasonably well-developed intuition. The goal has been to keep this essay short, using concepts of field theory to present the elements of energetics and dynamics in a unified manner.

For a treatment in detail of any particular item of elementary physics the reader should turn, as I did, to the magnificent lecture series of Feynman. Also listed at the end of this chapter is the historical account by Whittaker (1973) of the evolution of field theory. There is much good physics in those volumes. A powerful analysis of the fundamental problem of measurement and representation is offered in the book by Rosen (1978).

Considering its scope this is a short essay. But it is twice as long as one might think. It is written to be read twice. From its inception, this essay on complex systems has resisted serial organization. Overwhelmed by the subject's inherent complexity, I found that any concept I presented depended not only upon material on previous pages but also on material on the following pages. So there I sat, like Kekulé's snake, with my tail in my mouth.

By necessity, the straight line had to be abandoned for the circle. The chapter had to be written in cyclic form. Therefore, the reader is advised to read the chapter with confidence that all will be resolved upon completing the circle. Admittedly, that takes a certain amount of faith. On a second cycle, the reader, now conversant with an unfamiliar notation and having an overview of the scheme, will perceive, I hope, some merit in this unified approach to the physics of interacting systems. This caution applies especially to Section II, where the formalism to be used is laid out without benefit of elaboration or example. The following sections put flesh on these bare bones.

II. HIERARCHICAL STRUCTURE

A. Hierarchy

The hierarchical structure of a given system is the collection of physical descriptions (or representations) of that system as viewed from different levels of resolution. Whereas in relativity theory a system is viewed from different frames of reference in space–time, in hierarchy theory the various levels of description partake of the same space–time coordinates.

B. Levels of Description

The division into levels of description proceeds by aggregation, whereby a particle at one level is composed of a collection of particles of the next lower level of the hierarchy. The levels are ordered, and one can ascend or descend in the structure step-by-step as if on a ladder. In principle the hierarchy may be infinitely divisible into levels or may terminate in a lowest level composed of elementary monads. The hierarchy has a limit at the top, where the system as a whole is the sole unit, individual particles having lost their identities and contributing to the description by virtue of common, inherent properties. This is the systemic level. Other levels are called particulate levels.

C. Description Spaces

The analysis of the system starts with a choice of the level of the hierarchy at which particulate behavior will be described. This sets a description space at a given particulate level that shall be called the p-space. The description space at the systemic level shall be called the s-space. The analysis is based upon the physics of open, dissipative systems and is formulated in the continuum. The analysis of local behavior proceeds from the specification of particles and fluxes, the defining of potentials and forces, and the framing of equations of continuity. These are not sufficient to derive a complete physics of the system. Hierarchical considerations based upon the necessary congruence between the descriptions in p-space and in s-space are required.

D. Interactions

In space and time the system is permeated by a set of potential functions and their related fields. These potentials and fields provide a quantitative, local measure of the interaction of a unitary (test) particle with all other particles in the cosmos. In this Machian context the potentials and fields are state variables in s-space. The fields govern the dynamics and hence the fate of the system. Dynamics implies the motion of something, and as stated above it is how the system is divided into "things" that determines the level of the hierarchical structure under consideration.

E. Particles

The dynamics of a "thing," or rather of a particle as defined at a given level of physical description, proceeds from a specification of the forces it experiences. Dissipative systems (that is, complex real systems) are composed of a large number of particles, large enough to remove them from the realm of analytical mechanics. At any given particulate level in the hierarchy, the particles are grouped into classes—that is, into species, the members of each species having identical responses to each of the fields of interaction. A variable at the particulate level carries the subscript k to indicate that it has the value associated with species k. In this manner, the analysis is fixed at some level of the hierarchy. There will be no need for variables to carry an indicator identifying the chosen level.

F. Potentials

The potentials are functions of phase space, being evaluated at a position $\mathbf{R}$, a velocity $\mathbf{v}$, and a time t. There are two kinds of potentials, scalar potentials $\phi^i(\mathbf{R}, \mathbf{v} = 0, t)$, and vector potentials $\mathbf{\Phi}^i(\mathbf{R}, \mathbf{v}, t)$. The four-vector potential $[\phi^i; \mathbf{\Phi}^i]$, made by combining the two, will be called the potential. It is not required that all potentials have a vector part. When it is necessary to make a distinction, potentials that have a scalar and a vector part (such as electrical or gravitational) will be called real-potentials and those having only a scalar part (such as pressure or temperature) will be called pseudo-potentials. The potentials are systemic variables in that they characterize in a composite manner the mutual interactions of all the particles in the system. Real-potentials are calculated as retarded integrals over detailed interactions between particles, the intensity of interaction in one case (static part) being dependent upon the relative positions and in the other case (vector part) upon the relative motions of the particles. Pseudo-potentials arise because of the inadequacy of the description provided by a given p-space to encompass all the behavior manifested in the s-space description. Although it lacks detail, the s-space description is complete. In the p-space description, details that can only be resolved at a lower level of resolution must be incorporated, and this is done by "borrowing" some variables (pseudo-potentials) from s-space.

For much of the analysis of complex real systems it suffices to postulate the existence of the potentials without having to probe the nature of the field equations. The potentials and all other variables in s-space carry the superscript i.

G. Scalar Potential Energy

A particle of species k at the point $(\mathbf{R}, t)$ experiences the interactions characterized by the scalar potential, $\phi^i(\mathbf{R}, \mathbf{v} = 0, t)$ and possesses a potential energy, the magnitude depending upon a scale factor, A_k^i. The units of scalar potential energy are energy per mole of species k. At this point in the analysis, the particles have been identified and hence can be enumerated. No mention has yet been made of their physical properties, and so the alternative conventional units of energy per gram of species k will not be used.

The scalar potential energy possessed per mole of species k is

$$u_k(\mathbf{R}, t) = A^i_k \phi^i, \tag{1}$$

where the Einstein convention for summation over repeated indices is used. The doubly indexed scale factors establish a relationship between the systemic variables ϕ^i and the particulate variables u_k. Thus the scale factors provide a mapping from s-space to p-space. This mapping, which bridges the particle–interaction (or particle–field or wave–particle) duality, plays a central role in the investigation of complex real systems. The scale factors are not necessarily constants and may be functions of the potentials.

H. Vector Potential Energy

Likewise, there is a vector potential energy given by

$$\mathbf{U}_k(\mathbf{R}, \mathbf{v}, t) = A^i_k \mathbf{\Phi}^i. \tag{2}$$

It is assumed that scalar and vector potentials of the same index have identical scale factors; otherwise there would be bifurcations in mappings from s-space to p-space. Although it does not actually carry the units of energy, the variable $\mathbf{U}_k$ is called such because of its formal analogy to u_k. It has the units of momentum per mole of species k and will also be called momentum.

I. Driving Forces

Scalar potential energy is a measure of interactions whose intensity depends upon the relative positions of the universe of particles. The response to the change in interaction prompted by a change in the position of a particle is experienced locally as a force. The commonplace observation that things flow downhill is reflected in the definition of a driving force as the negative gradient of scalar potential energy:

$$\mathbf{F}_k(u) = -\nabla u_k. \tag{3}$$

J. Inertial Forces

Vector potential energy (momentum) is a measure of interactions whose intensity depends upon the relative motions of the universe of particles. Any change in the state of the motions is experienced locally as an inertial force defined as

$$\mathbf{F}_k(\mathbf{U}) = -D_{t,k}\mathbf{U}_k. \tag{4}$$

$D_{t,k}$ is the substantial time derivative with respect to the particle velocity $\mathbf{v}_k$ and is given as

$$D_{t,k} = \partial/\partial t + (\mathbf{v}_k \cdot \nabla). \tag{5}$$

Just as the driving force derives from a nonuniformity in space of the scalar potential energy, the inertial force derives from variations of the vector potential energy in time. The first term of the substantial derivative gives the local response to changes in interaction caused by variations in motion elsewhere. The second term is a measure of variations of the vector potential energy along the particle trajectory.

K. Net Force

The net force per mole of species k is given by the sum of the driving force and the inertial force. The net force can be represented by a four-vector "gradient" operator upon a four-vector potential energy:

$$\mathbf{F}_k = \mathbf{F}_k(u) + \mathbf{F}_k(\mathbf{U}) = -[\nabla; D_{t,k}][u_k; \mathbf{U}_k]. \tag{6}$$

The net force is the force available to overcome resistive (dissipative) forces, most often classed frictional or viscous. These resistive forces represent forces that can only be resolved at lower levels in the hierarchy. Dynamics is framed by specifying the net forces and the resistive forces in detail and equating them. If the resistive force (or equivalently the net force) is equal to zero, then the system is called ideal, nondissipative, or reversible. On the other hand, energetics aims at discovering the general behavior of the system—behavior not dependent upon a specification of the constitution of a given continuum. The use of net force in energetics allows a formulation free from models for resistive forces and the associated constitutive parameters.

L. Fields

The variables in the s-space description that correspond to forces in p-space are the fields. There is a field associated with spatial variations of the scalar potentials:

$$\Xi^i(\phi) = -\nabla\phi^i. \tag{7}$$

Likewise, there is a field associated with time variation along any given velocity:

$$\Xi^i(\mathbf{\Phi}) = -D_t\mathbf{\Phi}^i. \tag{8}$$

The potentials are defined in phase space, and the substantial derivative D_t is along the $\mathbf{v}$ for that point of phase space for which (7) and (8) are determined.

As in the case with forces, a net field is defined by a four-vector formulation:

$$\Xi^i(\mathbf{R}, \mathbf{v}, t) = -[\nabla; D_t][\phi^i; \mathbf{\Phi}^i]. \tag{9}$$

A basic premise of this analysis of complex real systems is that the scale factors are universal mappings between s-space and p-space representations. Therefore, the mapping between fields and forces must necessarily be

$$\mathbf{F}_k(\mathbf{R}, \mathbf{v}_k, t) = A^i_k \Xi^i(\mathbf{R}, \mathbf{v} = \mathbf{v}_k, t) \tag{10}$$

with similar expressions for the separate fields (7) and (8).

M. Properties

The motion described at the particulate level is the motion of particles. However, a particle carries with it more than just its identity. Each particle carries certain observable properties. These properties have a conjugate relationship to the potentials and equal them in number. The scale factor A^i_k is the measure of the amount of property associated with potential i that is carried per mole of species k. The total amount of property i, given as a density at a point in the continuum, is

$$a^i = A^i_k C_k, \tag{11}$$

where C_k is the local concentration of species k, given in units of moles per unit volume.

The concentrations C_k are the sole properties at the particulate level. This property is not so much physical as it is informational. Identity has been conferred upon things by forming categories of responses to forces. It is seen in (11) that the scale factors map properties in p-space to properties in s-space.

N. Fluxes

It is assumed that all members of a given species partake of the same motion. This is expressed analytically by prescribing a common velocity, $\mathbf{v}_k$. The velocities form a linearly independent set $\{\mathbf{v}_k\}$; that is, no one velocity can be expressed as a linear combination of the others. The velocities are given relative to a reference frame that does not move in response to forces impressed upon the system. In fact, the viewpoint is strictly Machian, with an inertial frame set in "the fixed stars" (remote galaxies).

The flux of species k is

$$\mathbf{J}_k = C_k \mathbf{v}_k. \tag{12}$$

It should be noted that the Einstein summation convention is not used in the preceding definition. Confusion is avoided because of the appearance of the subscript on the left-hand side; it is obvious that there is no summation. The flux of particles is mapped into s-space by the scale factors to give the flux of properties:

$$\mathbf{J}^i = A^i_k \mathbf{J}_k. \tag{13}$$

O. Continuity Equations

Particles are accountable. Their appearance and disappearance in a unit volume is not capricious but follows an equation of continuity:

$$\partial C_k/\partial t = -\nabla \cdot \mathbf{J}_k + \sigma_k. \tag{14}$$

The source term σ_k allows for, as it were, general creation and annihilation processes.

Properties are also accountable and hence obey equations of continuity:

$$\partial a^i/\partial t = -\nabla \cdot J^i + \sigma^i, \tag{15}$$

where σ^i is the source term in s-space.

P. Total Potential Energy

The total scalar potential energy carried by all the particles, given as a local density, is

$$u = C_k u_k. \tag{16}$$

The potential energy u is not a property. It has no equation of continuity *per se*. Or rather, one might say that its equation of growth $(\partial u/\partial t)$ is provided by an observation holding universally for all systems: the first law of thermodynamics.

The total vector potential energy (total momentum) carried by all the properties, given as a local density, is

$$\mathbf{U} = C_k \mathbf{U}_k. \tag{17}$$

The potential energy $\mathbf{U}$ is not a property and has no continuity equation. Its growth equation $(\partial \mathbf{U}/\partial t)$ describes the dynamics of the system.

III. FORCES AND FIELDS

A. Driving Forces

The idea of force as the negative gradient of potential energy is familiar even in elementary physics. In simple electrostatics, for example, the potential at $\mathbf{R}$

generated by a point charge q at the origin is

$$\phi(\mathbf{R}) = q/4\pi\varepsilon_0 R. \tag{18}$$

The potential energy of a second point charge Q inserted in the potential at $\mathbf{R}$ is

$$u(\mathbf{R}) = Q\phi(\mathbf{R}). \tag{19}$$

The electric field is

$$\mathbf{E} = -\nabla\phi \tag{20}$$

and the force is

$$\mathbf{F} = Q\mathbf{E}. \tag{21}$$

The electric charge Q is the scale factor that maps potential to potential energy and field to force. In a similar manner, the force of gravity is calculated as the gradient of a potential function, with mass as the scale factor.

The situation is completely analogous in complex real systems even though the potentials are retarded so as to take into account the finite speed of modes of interaction. Furthermore, the analogy is not disturbed even though the analysis is done in the continuum. As a matter of fact, it is in the continuum that one sees that gradients of pseudo-potentials give rise to forces. Fourier's law states that the gradient of temperature is the force that drives the flow of thermal energy. Poiseuille's equation describes laminar volume flow driven by the force of a pressure gradient.

The forces and scale factors associated with the level of the hierarchy wherein the world is seen as molecules (a word taken here to include ions) are well known in physical chemistry. To give the intuition something concrete to grasp, Table I lists these familiar potentials and their associated scale factors.

Table I

Potentials and Associated Scale Factors

Potential	Scale Factor
Electric, ψ_e	Partial molar charge, $\bar{Z}_k$
Gravitational, ψ_g	Partial molar mass, $\bar{M}_k$
Pressure, p	Partial molar volume, $\bar{V}_k$
Thermal, T	Partial molar entropy, $\bar{S}_k$
Informational,[a] μ_k	Unity, 1

[a] Concentration-dependent part of the Gibbs potential.

However, it is only in later chapters that concepts such as entropy and informational potentials are developed.

B. Forces in the Continuum

The scale factors map fields to forces per mole of species k. In the continuum one examines the dynamics of a unit volume. Forces per mole are converted to forces per unit volume upon multiplication by the local molar concentration C_k, with the total force density being given by the sum over k. For example, the driving force density is

$$\mathbf{F}(u) = C_k \mathbf{F}_k(u) \tag{22}$$

with similar sums for the vector force and the net force.

The preceding equation is the representation of force density in p-space. The corresponding representation in s-space is found by writing $\mathbf{F}(u)$ explicitly as a mapping from the fields. This gives

$$\mathbf{F}(u) = C_k[A_k^i \Xi^i(\phi)] = a^i \Xi^i(\phi) = \mathbf{F}(\phi) \tag{23}$$

under the definition (11) of property. The p-space and s-space representations are identical in form. As will be seen, this result applies to a class of bilinear forms and plays an important role in the analysis of complex real systems. In warrants the stating of the following theorem.

THEOREM. There is a congruence between the representation of the driving force densities in p-space and in s-space. The form of the representation is a bilinear invariant

$$C_k \mathbf{F}_k(u) = a^i \Xi^i(\phi). \tag{24}$$

C. Dynamics

The formulation of dynamics is formally simple. The net force is set equal to the resistive force. Conventionally, the resistive forces are characterized by the divergence of the stress tensor $\mathbf{T}$. The stress tensor depends on the constitutive properties of the continuum. Apart from an admission of its existence, little else will be required of the stress tensor in this section. It is discussed in detail in Section III and in a wealth of excellent books on continuum mechanics. In the order of increasing explicitness, the dynamical equation can be written

$$\begin{aligned} \mathbf{F} &= \nabla \cdot \mathbf{T} \\ C_k \mathbf{F}_k &= \nabla \cdot \mathbf{T} \\ -C_k A_k^i[\nabla; D_{t,k}][\phi^i; \mathbf{\Phi}^i] &= \nabla \cdot \mathbf{T}. \end{aligned} \tag{25}$$

Note that one must not hastily sum over k in the last expression and put $a^i = C_k A_k^i$ in front of the gradient. The vector potentials and the vector fields are functions of phase space: that is, $\mathbf{\Phi}^i(\mathbf{R}, \mathbf{v}, t)$ and $\mathbf{\Xi}^i(\mathbf{R}, \mathbf{v}, t)$. When one uses a scale factor to map them to the vector potential energy of and the inertial force on a given species k, one must carefully note that $\mathbf{v}$ is replaced by $\mathbf{v}_k$. Also, upon this mapping from s-space to p-space, $D_t \to D_{t,k}$. Thus in (25), $D_{t,k}$ operates upon $\mathbf{\Phi}^i(\mathbf{R}, \mathbf{v} = \mathbf{v}_k, t)$, and the summation over k on the left-hand side is actually over three terms. This argument, by the way, shows that the preceding theorem cannot be generalized to inertial forces. Certainly, there must be a congruence with the representation of the inertial force densities in p-space and in s-space. However, the congruence does not give rise to a bilinear invariant because the injection $\mathbf{v} \to \mathbf{v}_k$ makes a mapping that obviously is not bilinear. That is, the mapping $\mathbf{F}_k(\mathbf{U}) = A_k^i \mathbf{\Xi}^i(\mathbf{\Phi})$ actually leads to $F(\mathbf{U}) = C_k F_k(\mathbf{U}) = \mathrm{C}_k[A_k^i D_{t,k} \mathbf{\Phi}^i(\mathbf{R}, \mathbf{v} = \mathbf{v}_k, t)]$.

After a brief discussion of the substantial derivative, it will be shown by a comparison to fluid dynamics and electrodynamics that the general dynamical equation (25) encompasses the phenomena of gravitational inertia and electrical inertia (magnetic field). A thorough examination of the general dynamical equation is presented in Section VI.

D. Substantial Derivative

The substantial derivative of a scalar function $f(x, y, z, t)$ along a trajectory determined by a velocity $\mathbf{v} = (dx/dt, dy/dt, dz/dt)$ is

$$D_t f = \frac{\partial f}{\partial t} + \frac{\partial f}{\partial x}\frac{dx}{dt} + \frac{\partial f}{\partial y}\frac{dy}{dt} + \frac{\partial f}{\partial z}\frac{dz}{dt} \tag{26}$$

or in vector notation

$$D_t f = \partial f/\partial t + \mathbf{v} \cdot \nabla f. \tag{27}$$

The substantial derivative of a vector function $\mathbf{\Phi}(x, y, z, t)$ along the velocity is formally written

$$D_t \mathbf{\Phi} = \partial \mathbf{\Phi}/\partial t + (\mathbf{v} \cdot \nabla)\mathbf{\Phi}. \tag{28}$$

This equation bears a pleasing resemblance to (27) but is not too useful in calculations. An improvement is to write the last term as the dyad $\nabla\mathbf{\Phi}$ dotted with $\mathbf{v}$; that is,

$$D_t \mathbf{\Phi} = \partial \mathbf{\Phi}/\partial t + \mathbf{v} \cdot \nabla\mathbf{\Phi}. \tag{29}$$

A slight change in notation will allow vector operators to be used in vector identities. This in turn will allow the substantial derivative of the vector potential to be cast into a form that, by comparison with the fundamental equations of electrodynamics and continuum mechanics, shows that it indeed represent inertial forces. By convention, $\mathbf{\nabla}$ operates upon everything to its right, and this causes trouble when $\mathbf{\nabla}$ is moved about in vector identities. Therefore, to keep in mind what $\mathbf{\nabla}$ operates on, the following notation is introduced:

$$(\mathbf{v} \cdot \mathbf{\nabla})\mathbf{\Phi} \equiv (\mathbf{v} \cdot \mathbf{\nabla}_{\mathbf{\Phi}})\mathbf{\Phi}. \tag{30}$$

Substituting $\mathbf{v}$, $\mathbf{\nabla}_{\mathbf{\Phi}}$, and $\mathbf{\Phi}$ into the identity

$$\mathbf{A}(\mathbf{B} \cdot \mathbf{C}) = \mathbf{C}(\mathbf{B} \cdot \mathbf{A}) + \mathbf{B} \times (\mathbf{A} \times \mathbf{C}) \tag{31}$$

allows the substantial derivative to be written in vector notation as

$$D_t\mathbf{\Phi} = \partial\mathbf{\Phi}/\partial t - \mathbf{v} \times (\mathbf{\nabla} \times \mathbf{\Phi}) + \mathbf{\nabla}_{\mathbf{\Phi}}(\mathbf{v} \cdot \mathbf{\Phi}), \tag{32}$$

where $\mathbf{\nabla}_{\mathbf{\Phi}}$ operates only upon the components of $\mathbf{\Phi}$.

E. Fluid Dynamics

For the situation where gravity and pressure provide the driving forces, the dynamics of a simple fluid is governed by the familiar equation

$$-\rho D_t\mathbf{v} - \rho\mathbf{\nabla}\psi_{\mathrm{g}} - \mathbf{\nabla}p = \mathbf{\nabla} \cdot \mathsf{T}, \tag{33}$$

where ρ is the local mass density. In four-vector notation, this can be written

$$-\rho[\mathbf{\nabla}; D_t][\psi_{\mathrm{g}}; \mathbf{v}] - [\mathbf{\nabla}; D_t][p; 0] = \mathbf{\nabla} \cdot \mathsf{T}. \tag{34}$$

In a simple fluid all particles move with a common velocity; and hence the inertial field is the same for all; namely, $D_{t,k}\mathbf{\Phi}^i(\mathbf{R}, \mathbf{v} = \mathbf{v}_k, t) = D_t\mathbf{\Phi}^i(\mathbf{R}, \mathbf{v}, t)$. This means that one can sum over k in the general dynamics (25) to obtain $a^i = A^i_k C_k$ as the coefficient for the four-vector gradient. A comparison of (34) and (25) leads to the following conclusions: (1) the left-hand side of (34) is of the form $a^i \Xi^i$, (2) the property associated with the gravitational field is mass, (3) the property associated with the pressure field is unity, being volume per unit volume, (4) the pressure field has no vector potential, and (5) the gravitational vector potential is $\mathbf{\Phi}_{\mathrm{g}} = \mathbf{v}$. As will be seen in Section VI, this identification of the gravitational vector potential with the particle velocity (in an inertial frame) is the result of a retarded integral of mass fluxes over the cosmos. In this Machian viewpoint, inertia arises because of a change in motion relative to the "fixed stars."

Consider for a moment the gravitational inertia term of (34):

$$D_t\mathbf{\Phi}_{\mathrm{g}} = \partial\mathbf{\Phi}_{\mathrm{g}}/\partial t + (\mathbf{v} \cdot \mathbf{\nabla})\mathbf{\Phi}_{\mathrm{g}} \tag{35}$$

with $\mathbf{\Phi}_g = \mathbf{v}$. By (32) the final term of this equation can be written

$$(\mathbf{v} \cdot \nabla)\mathbf{\Phi}_g = (\nabla \times \mathbf{\Phi}_g) \times \mathbf{v} + \nabla_{\mathbf{\Phi}_g}(\mathbf{v} \cdot \mathbf{\Phi}_g), \tag{36}$$

where a sign change occurs from a reversing of the cross product. Consider now the last term of the preceding equation. Because of the symmetry introduced upon substitution of $\mathbf{\Phi}_g = \mathbf{v}$ into the parentheses, this term becomes

$$\nabla_{\mathbf{\Phi}_g}(\mathbf{v} \cdot \mathbf{\Phi}_g) = \tfrac{1}{2}\nabla(\mathbf{v} \cdot \mathbf{v}). \tag{37}$$

The inertial force of fluid dynamics can therefore be written in the familiar form

$$D_t\mathbf{\Phi}_g = \partial\mathbf{v}/\partial t + \mathbf{\Omega} \times \mathbf{v} + \tfrac{1}{2}\nabla v^2, \tag{38}$$

where $\mathbf{\Omega} = \nabla \times \mathbf{v}$ is defined as the vorticity.

F. Electrodynamics

The net force per mole of an ionic species having a partial molar change $\bar{Z}_k$ is

$$F_k = A_k^i \Xi^i(\mathbf{R}, \mathbf{v} = \mathbf{v}_k, t) = -\bar{Z}_k[\nabla; D_{t,k}][\psi_e; \mathbf{A}], \tag{39}$$

where only the electric potentials are considered, with the standard convention of ψ_e denoting the scalar potential and $\mathbf{A}$ the vector potential. This force is traditionally called the ponderomotive force and by (32) can be expanded to

$$\mathbf{F}_k = -\bar{Z}_k\nabla\psi_e - \bar{Z}_k\,\partial\mathbf{A}/\partial t + \bar{Z}_k\mathbf{v}_k \times (\nabla \times \mathbf{A}) - \bar{Z}_k\nabla_{\mathbf{A}}(\mathbf{v}_k \cdot \mathbf{A}). \tag{40}$$

Were it not for the last term, this would look like the familiar Lorentz force of elementary electrodynamics. However, the elementary treatments of the ponderomotive force are done in a fixed system of coordinates with, in particular, the vector potential as a retarded integral over currents relative to the origin. Lorentz himself (1904) derived the ponderomotive force in moving coordinates, obtaining (40). His vector potential in moving coordinates was in terms of currents relative to the motion of the charge under consideration, this is, currents relative to $\mathbf{v}_k$. Even though this relativistic nature of $\mathbf{A}$ is not explicitly indicated above, it should be kept in mind. In Section II it was stressed that the vector potentials $\mathbf{\Phi}^i(\mathbf{R}, \mathbf{v}, t)$ and vector fields $\Xi^i(\mathbf{R}, \mathbf{v}, t)$ are defined in phase space; and in the mapping to the energies and forces identified with particles, $\mathbf{v}$ is set to $\mathbf{v}_k$. A more complete discussion of the Lorentz force is given in Section VI. The goal of the present section was to show simply that the four-vector gradient operating upon the four-vector potential gives the force term of Lorentz's magneto-electrodynamics.

G. Dynamics at the Molecular Level

The potentials and scale factors at the molecular level of description are listed in Table I. At this level the dynamics (25) is

$$C_k\bar{M}_k[\nabla; D_{t,k}][\psi_g; \mathbf{v}_k] + C_k\bar{Z}_k[\nabla; D_{t,k}][\psi_e; \mathbf{A}(\mathbf{v}_k)] + V_I[\nabla; D_{t,k}][p; 0]$$
$$+ s[\nabla; D_{t,k}][T; 0] + C_k[\nabla; D_{t,k}][u_k; 0] = -\nabla \cdot \mathbf{T}. \quad (41)$$

This equation is put in very explicit terms for illustrative purposes. Even though mass density is $\rho = C_k\bar{M}_k$ and charge density is $z = C_k\bar{Z}_k$, these sums cannot be performed above; the vector potentials depend upon relative motion and hence upon $\mathbf{v}_k$. To show the volume is the property associated with the pressure potential, a unity volume density $V_I = C_k\bar{V}_k = 1$ is defined. The quantity V_I is a density, being total volume per unit volume. The property associated with the temperature (thermal) potential is entropy, $s = C_k\bar{S}_k$. Finally, as noted before, the scale factor associated with the informational potential μ_k is the identity, being moles per mole. For the sake of a unified notation, the pseudo-potentials are given as four vectors with zero vector potentials.

IV. CONGRUENCE OF REPRESENTATIONS

A. Perpetual Motion

Perpetual motion of the first kind is prohibited. That is, it is impossible to extract an increment of work from the system upon traversing a closed loop. However, the systems considered here are dynamically open; so consideration of work integrals must account for the fact that energy can be "pumped" in by external agencies that vary the fields in time. Therefore, the work integral must be performed statically, that is, with time frozen, as it were, at some instant $t = \tau$. The dictum against perpetual motion is then

$$\oint \mathbf{F}_k[u(\tau)] \cdot d\mathbf{s} = 0. \quad (42)$$

B. Conservative Forces

The necessary and sufficient condition to satisfy (42) is that $\mathbf{F}_k(u)$ be the gradient of a scalar function. And it was postulated in Section II that the driving forces are so constituted that things tend to flow downhill; that is,

$$\mathbf{F}_k(u) = -\nabla u_k. \quad (43)$$

This, by the way, does not imply that the system is conservative in the sense that dissipation is not allowed. Nor does it imply that work cannot be extracted in time around a closed loop; an electric motor, for example, performs cyclic work by virtue of the mains voltage, which constantly varies what is locally downhill.

C. Conservative Fields

As far as the p-space representation is concerned, the postulate of gradient forces is equivalent to the stricture against perpetual motion. What about in s-space? The fields are given as

$$\Xi^i(\phi) = -\nabla\phi^i \tag{44}$$

and are mapped to forces in p-space via

$$\mathbf{F}_k(\phi) = A^i_k\Xi^i(\phi) = -A^i_k(\phi)\nabla\phi^i. \tag{45}$$

The possible dependence of the scale factors upon the potentials is expressly indicated above. If indeed that were the case, then the preceding mapping between the two representations gives a force that apparently is not the gradient of some scalar function.

Does (45) violate the injunction against perpetual motion stated in (42) or put any conditions upon the scale factors so that (42) is assured? No. The conservative constraint (42) states that the integral

$$\oint A^i_k(\phi)\nabla\phi^i \cdot d\mathbf{s} \tag{46}$$

must be zero. This line integral is immediately reduced to quadratures via the identity

$$A^i_k(\phi)\nabla\phi^i \cdot d\mathbf{s} = A^i_k(\phi)\,d\phi^i. \tag{47}$$

The space variables disappear, and the integral (46) is zero since the potentials ϕ^i are independent of each other along any path. The mapping from fields to forces is such that the system representation remains statically conservative.

On the other hand, potentials are mapped to potential energies. Does this put any conditions upon possible mappings so that force representations† will be consistent? That is, what is the consequence of

$$\mathbf{F}_k = -\nabla u_k = -\nabla A^i_k\phi^i \tag{48}$$

† For convenience that explicit functional dependence of $\mathbf{F}_k(u)$ and $\mathbf{F}_k(\phi)$ will be suppressed in the following calculations.

standing in contrast to

$$\mathbf{F}_k = A_k^i \Xi^i = -A_k^i \nabla \phi^i. \tag{49}$$

Since the scale factors may depend upon the potentials, the gradient in (48) expands to

$$\mathbf{F}_k = -A_k^i \nabla \phi^i - \phi^i (\partial A_k^i / \partial \phi^j) \nabla \phi^j. \tag{50}$$

The indices i and j can be interchanged to give

$$\mathbf{F}_k = -[A_k^i + \phi^j \, \partial A_k^j / \partial \phi^i] \nabla \phi^i. \tag{51}$$

A comparison with (49) shows that the requirement of consistent representations necessitates the following theorem:

THEOREM. Any functional dependence of the scale factors upon the potentials must satisfy

$$\phi^j \, \partial A_k^j / \partial \phi^i = 0 \qquad \text{for any } i, j, \text{ and } k. \tag{52}$$

COROLLARY. If it is established that there is one scale factor that is a function of the potentials, then there must be at least two scale factors that are functions of the potentials.

The preceding considerations can be summarized by the statement that the mapping A_k^i and the gradient operator ∇ are commutative. Figure 1 is a commutativity diagram for the two operations.

D. Accountability

The preceding theorem has an important bearing upon the relationship between the equations of continuity in p-space and in s-space. This however, is

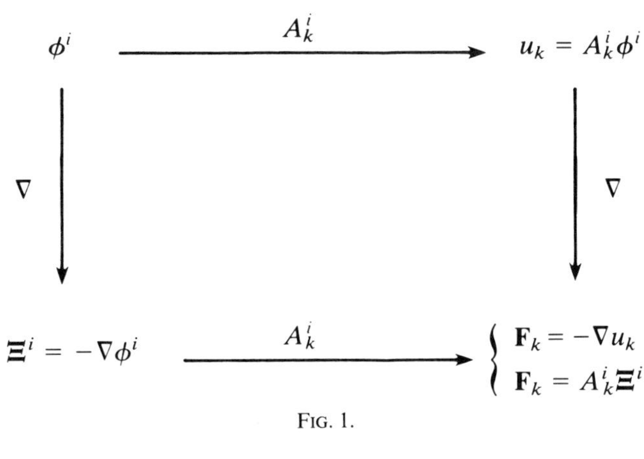

FIG. 1.

not immediately evident. There is a congruence that starts with, but is not resolved by, the idea of accountability of properties stated in Section II.O. Based upon Gauss's theorem, the equation of continuity presents a canonical form for accounting that must be preserved in both representation spaces. In p-space the sole property is particle identity and is measured by concentration; thus we have the accounting

$$\partial C_k/\partial t = -\nabla \cdot \mathbf{J}_k + \sigma_k, \tag{53}$$

where σ_k is the source term. Likewise, the canonical accounting for systemic properties is

$$\partial a^i/\partial t = -\nabla \cdot \mathbf{J}^i + \sigma^i, \tag{54}$$

where σ^i is the source term for properties in generalized creation and annihilation processes.

The mappings $C_k \to a^i$ and $\mathbf{J}_k \to \mathbf{J}^i$ by means of the scale factors allow the derivation of the systemic equation (54) from the particulate equation (53). One obtains

$$\frac{\partial a^i}{\partial t} = -\nabla \cdot \mathbf{J}^i + A^i_k\sigma_k + \mathbf{J}_k \cdot \frac{\partial A^i_k}{\partial \phi^j}\nabla\phi^j + C_k\frac{\partial A^i_k}{\partial \phi^j}\frac{\partial \phi^j}{\partial t}. \tag{55}$$

A comparison of (55) with (54) shows that the source term σ^i is not given by the mapping $A^i_k\sigma_k$. There are other terms. The equations of continuity are not mapped in a simple manner in passing from the representation in p-space to that in s-space.

Although there is a congruence in form between the equations of continuity, the relationship between them (55) derived from the mapping of C_k and $\mathbf{J}_k$ to s-space is neither simple, elegant, nor invested with immediately apparent physical meaning. There is, nevertheless, a physically meaningful congruence springing from the accountability inherent in the equations of continuity that becomes evident if attention is shifted from properties to energy. The source term σ_k is the rate at which newly created particles enter the continuum (or leave, if σ_k is negative). Each new particle has associated with it an energy u_k. This is just the potential energy it possesses by virtue of appearing in the continuum permeated by the set of local potentials ϕ^i. Thus the rate at which this energy appears is $u_k\sigma_k$. From the point of view of the s-space representation, properties are appearing at the rate of σ^i, and the rate at which energy appears is $\phi^i\sigma^i$. Physical intuition deems it imperative that the energy source as represented in p-space must equal that as represented in s-space.

Although this equivalence seems self-evident in that it precludes appearance of energy sources in one representation not evident in the other, it actually is a consequence of the prohibition of perpetual motion of the first kind. The systemic source term as derived from the mapping of C_k and $\mathbf{J}_k$ to s-space is given by the last three terms of (55). The rate of the appearance of energy in s-space can thus be written

$$\phi^i\sigma^i = \phi^i A^i_k\sigma_k + \mathbf{J}_k \cdot \phi^i \frac{\partial A^i_k}{\partial \phi^j}\nabla\phi^j + C_k\phi^i \frac{\partial A^i_k}{\partial \phi^j}\frac{\partial \phi^j}{\partial t}. \tag{56}$$

If the summation over i is made first, it is immediately seen that by Theorem (52) the last two terms of the preceding equation are zero. This establishes an important congruence that plays an essential role in the derivation of the equation of continuity for entropy to be presented in the following chapter. This result is similar to that stated in Theorem (24) and merits being stated as a theorem.

THEOREM. There is a congruence between the representations of energy source terms in p-space and in s-space. The representation is an invariant bilinear form:

$$\phi^i\sigma^i = u_k\sigma_k. \tag{57}$$

Finally it must be pointed out that the energy source terms in the preceding discussion apply to those energies invested in a particle by virtue of its being in a region of potential ϕ^i. It does not encompass the energy released from bonds of aggregation (or absorbed, as the case may be) as a particle changes its internal structure to give up component particles or to accept additional component particles. The subsequent rearrangement of bonding forces may release energy to the system or absorb energy. Examples include fusion, fission, chemical reactions, and radioactive decay. The accounting of such energies takes place within the context of the first law of thermodynamics, the theme of the following section.

E. Another Congruence

An amount of potential energy u_k flows by convection with the particle flux $\mathbf{J}_k$. If there is a divergence in particle flux, there will be an increase or decrease of energy in the unit volume such that

$$\text{energy change due to convection} = -u_k\nabla \cdot \mathbf{J}_k. \tag{58}$$

The corresponding expression in s-space is $-\phi^i\nabla \cdot \mathbf{J}^i$. Physical intuition deems that these two representations agree. The mathematical proof of this depends upon the stricture against perpetual motion. By the mapping $\mathbf{J}_k \to \mathbf{J}^i$,

$$\nabla \cdot \mathbf{J}^i = \nabla \cdot A^i_k\mathbf{J}_k \tag{59}$$

which is, upon expansion,

$$\nabla \cdot \mathbf{J}^i = A^i_k\nabla \cdot \mathbf{J}_k + \mathbf{J}_k \cdot \frac{\partial A^i_k}{\partial \phi^j}\nabla\phi^j. \tag{60}$$

Multiplication by ϕ^i gives

$$\phi^i \nabla \cdot \mathbf{J}^i = \phi^i A_k^i \nabla \cdot \mathbf{J}_k + \mathbf{J}_k \cdot \phi^i \frac{\partial A_k^i}{\partial \phi^j} \nabla \phi^j. \tag{61}$$

The last term is zero by Theorem (52). Thus, under the mapping $\phi^i \to u_k$, the preceding equation becomes the following theorem.

THEOREM. There is a congruence between the representations of the local change in scalar potential energy due to convection as given in s-space and in p-space. It is given by the invariant bilinear form

$$u_k \nabla \cdot \mathbf{J}_k = \phi^i \nabla \cdot \mathbf{J}^i. \tag{62}$$

F. Scale Factors

In Section IV.D the accountability of the properties a^i was examined. This section ends with one more observation on the nature of properties, or, rather, the mapping of p-space properties C_k to s-space properties a^i. The mapping is

$$a^i = A_k^i C_k. \tag{63}$$

Note that if all concentrations are increased by the factor λ, all properties will also be increased by the factor λ. In other words, (63) is a homogeneous function of the first order in C_k. This needs some elaboration. It assumes that the scale factors are not functions of concentration—which is true. However, they may well be functions of potentials. Therefore, the preceding statement applies in the situation where the concentration changes cause no changes in the potentials. Euler's theorem for homogeneous functions applied to $a^i = a^i(\phi^j, C_k)$ gives

$$a^i = (\partial a^i / \partial C_k) C_k. \tag{64}$$

A comparison of (64) and (63) gives the following theorem.

THEOREM. The scale factors, which in general provide a mapping between the two representation spaces, can be calculated by the following fundamental relationship between properties in s-space and in p-space:

$$A_k^i = \partial a^i / \partial C_k. \tag{65}$$

V. SCALAR POTENTIAL ENERGY

A. Energy Conservation

The motions of particles are driven by forces directed down gradients of potentials. In real systems, motion is resisted by inertial and frictional forces. In ideal systems the only forces that resist motion are the inertial forces. A decrease in scalar potential energy during the motion is reflected in a corresponding increase in vector potential energy (momentum). In analytical mechanics this exchange is the content of Newton's second law:

$$d/dt(m\mathbf{v}) = -\nabla u. \tag{66}$$

A spatial variation in a scalar quantity is balanced by a temporal variation in a vector quantity. The absence of a frictional force term in (66) defines the system as ideal and allows an immediate integration giving a balance of quantities carrying the same units and of the same tensorial order. The loss of potential energy equals the gain in kinetic energy. Total energy is conserved.

The real world, on the other hand, is replete with situations where energy is not conserved. This, in fact, is what characterizes the "real." Energy may not be strictly conserved, but it is accountable. This accountability makes clear the necessity of a thermal potential energy. Casual observation makes obvious its existence. Analysis will reveal its behavior.

B. Temperature

A steel ball dropped into a vertical column of viscous oil will soon reach a steady-state motion. Its velocity becomes constant. It loses gravitational potential energy along its trajectory without, however, a compensating increase in kinetic energy. The gravitational energy thus degraded must be accounted for. Either it flows away or it stays around. Any energy that stays around raises the temperature of the system. Scalar potential energy of gravity is converted to scalar thermal energy.

As a second example, consider a block of wood sliding along a table. Some agency, which no longer acts upon it, has given it an initial velocity. The block loses kinetic energy but gains no potential energy in its motion along the horizontal table top. Vector potential energy is converted to scalar thermal energy.

From the preceding observations it must be concluded that among the set of potentials characterizing any real system there necessarily exists a scalar potential energy that is the repository of those energies degraded in motions. There is necessarily a scalar potential function associated with the potential energy. It will be called temperature and be denoted by T. In the enumeration of the potentials, temperature will be given the superscript m; thus, $T = \phi^m$. Like all other potentials, T is a function of space and time.

A potential energy and its underlying potential have been postulated. There must be a scale factor, A_k^m, such that

$$u_k^m = A_k^m T, \tag{67}$$

where u_k^m is the thermal energy per mole of species k. At the molecular level of description, A_k^m is $\bar{S}_k$, the partial molar entropy. For most of the following discussion this notation will be used because of its familiarity, although the results apply to any level of the hierarchy of descriptions.

C. Entropy

As with all other potentials, there is a systemic property conjugate to the thermal potential T. This property will be called entropy and will be denoted by $s = a^m$. Like all other systemic properties, it is a density and is obtained from the p-space properties (concentrations) by the mapping afforded by the scale factor A_k^m:

$$a^m = A_k^m C_k \tag{68}$$

or equivalently,

$$s = \bar{S}_k C_k. \tag{69}$$

Entropy is a property and, like all other properties, must have an equation of continuity. However, the equation for $s = a^m$ is not simply (54) with the index m. One knows from experience that, in addition to the entropy flux $\mathbf{J}^m$, thermal energy is carried by heat conduction $\mathbf{J}_q$. Furthermore, in addition to the entropy source term σ^m, there must be a source term σ to characterize thermal energies arising from motion. To retain the form of representation, an equation of continuity for entropy is given as

$$\partial s/\partial t = -\nabla \cdot \mathbf{J}^s + \sigma^s, \tag{70}$$

where $\mathbf{J}^s = \mathbf{J}^m + f(T, \mathbf{J}_q)$ and $\sigma^s = \sigma^m + \sigma$. The form has been preserved at the price of introducing two undetermined functions: $f(T, \mathbf{J}_q)$ and σ. At this point, the conjugate thermal variables temperature and entropy have been introduced in accord with the formalism established in Section II. Equation (70) has, however, reached the limits of formal analogy. The formalism must now be supplemented with physics.

D. Entropy Flux

Consider first the flux of entropy $\mathbf{J}^s$. As with all other properties, entropy is assigned to particles. Entropy will flow by convection, then, as particles move with their species velocities $\mathbf{v}_k$. With m as the superscript denoting thermal variables, the convectional flux of entropy, like any other property, is

$$\mathbf{J}^m = \bar{S}_k J_k. \tag{71}$$

Superimposed upon the coherent motions represented by the species velocities $\mathbf{v}_k$ are random motions. These motions also are a form of energy, but not quite as accessible for doing work as motions observed to be coherent at the given level of description in the hierarchy. Even at apparent rest ($\mathbf{v}_k = 0$), these motions persist, in effect, as minute agitations. These agitations are assigned to particles, and hence their energies are accountable within the formalism. These motions are random and therefore accountable as scalar energy. Their energy will be included in the thermal potential energy. If such energy of agitation is not distributed uniformly in space, it will flow by diffusion (momentum) transfer); obviously, this is the conducted heat $\mathbf{J}_q$ of elementary physics.

It remains to show how the heat flux $\mathbf{J}_q$ and the convectional flux of entropy $\mathbf{J}^m$ are related to the entropy flux $\mathbf{J}^s$. Energy units provide a common currency. Heat flux is already in energy units. Consider the total flux of scalar potential energy. As with properties, scalar potential energy is assigned to particles and hence moves (by convection) with particles. However, there is a motion of energy $\mathbf{J}_q$ independent of the coherent motion of particles. The representation scheme for property fluxes as a mapping $\mathbf{J}^i = A^i_k \mathbf{J}_k$ is not sufficient for energy fluxes. One must write

$$\text{energy flux} = u_k \mathbf{J}_k + \mathbf{J}_q. \tag{72}$$

The representation scheme of Section II is thus strained because of the need in real systems to include thermal effects. A particular level in the hierarchy has been chosen, and a p-space description of the system has been framed at that level. Thermal effects, however, are resolvable only at a lower level of the hierarchy. Only at a lower level would there be instruments of resolution fine enough to observe coherent motions or engines of a scale small enough to extract work in a direct manner. Nevertheless, the description of the system in the chosen p-space and in s-space must be complete. The formalism must be augmented to include these intrusions of effects belonging to lower levels in the hierarchy. This augmentation is implicit in the extra source term of (70), in the so far unexplained distinction between $\mathbf{J}^m$ and $\mathbf{J}^s$, and in (7.2). The resolution of this problem of representation hinges upon the fact that temperature is a potential. The description in s-space is constructed within the formalism with temperature as the sole thermal potential. This will define $\mathbf{J}^s$. The entropy production term σ will then be found by insisting upon a congruence between the p-space and the s-space descriptions of real (thermal) systems. The first law of thermodynamics, under the constraint of this congruence, gives σ as a function of state variables.

The energy flux equation can, by means of the mappings $\mathbf{J}_k \to \mathbf{J}^i$ and $\phi^i \to u_k$, be written in s-space as

$$\text{energy flux} = \phi^i \mathbf{J}^i + \mathbf{J}_q. \tag{73}$$

The entropy flux $\mathbf{J}^s$ will now be defined so that the representation of energy flux in s-space is the simple sum of property fluxes times potentials. The thermal potential is $\phi^m = T$, the thermal property is s, and the flux of thermal property is $\mathbf{J}^s$. So the flux of thermal energy must, under the formalism, by $T\mathbf{J}^s$. What is $\mathbf{J}^s$?

The thermal contribution to the summation in (73) can be separated out to give

$$\text{energy flux} = \sum_{i \neq m} \phi^i \mathbf{J}^i + \phi^m \mathbf{J}^m + \mathbf{J}_q. \tag{74}$$

The last two terms are the flux of thermal energy and must, to preserve the formalism in s-space, equal $T\mathbf{J}^s$. Therefore

$$\mathbf{J}^s = \mathbf{J}^m + (1/T)\mathbf{J}_q. \tag{75}$$

This preservation of form is central to the concept of entropy. To emphasize that the representation of energy flux in s-space is now, by construction, a bilinear function of potentials and property fluxes, it shall be written explicitly as:

$$\text{energy flux} = \sum_{i \neq m} \phi^i \mathbf{J}^i + T\mathbf{J}^s. \tag{76}$$

E. Entropy Source Terms

The equation of continuity for entropy (70) has two source terms—an admission that the representation scheme must be augmented. The source term σ^m is like the other source terms σ^i for properties. It represents the appearance (or disappearance) of entropy associated with creation and annihilation processes, each particle in the process bearing the entropy per mole of $A_k^m = \bar{S}_k$. In the discussions in Section IV on sources, it was, of course, assumed that the representations were complete, that all properties were included in the summations. Even though the thermal variables (indexed by m) were not singled out for examination until Section V, they were implicitly included in the summations in the preceding chapters.

The source term σ in (70) does not, however, carry the superscript m. It stands outside the heretofore developed formalism. Entropy is the systemic property associated with the thermal potential $\phi^m = T$. As seen in preceding examples, thermal energy is created during motion, being as it were the repository of degraded scalar and/or vector potential energies. The representational formalism will be used to find an analytical expression for any sources of entropy that might arise from the motion. This is not the only possibly way in which entropy might enter the representation. Observation shows that thermal energy may be released or absorbed during a chemical reaction. This will also, no doubt, have to be part of the source term σ, not being included in the accounting provided by the source term σ^m.

F. Chemical Reactions

In the case of chemical reactions, a local DeDonder reaction rate $d\xi_R/dt$ can be written for each reaction R. The source term for molecules of species k is then

$$\sigma_k = \nu_{kR}\, d\xi_R/dt \tag{77}$$

where the ν_{kR} are the stoichiometric coefficients, being positive for products and negative for reactants. The summation convention applies to R so that (77) is the source for all reactions.

The source term for properties is not immediately derivable from the particle source term by a mapping. This problem was addressed in Section IV. For example, if the scale factors A_k^i are functions of potentials, then a^i can change in time even though there are no flows or reactions. Thus even though $\partial C_k/\partial t = 0$,

$$\partial a^i/\partial t = (\partial/\partial t)A_k^i C_k = C_k(\partial A_k^i/\partial\phi^j)(\partial\phi^j/\partial t), \tag{78}$$

which will not be zero if some potentials vary in time. Therefore a^i may have a source term even when all $\sigma_k = 0$.

Consider that part of σ^i that does arise because of chemical reactions; denote it by σ_R^i. The reaction takes place at a point in the continuum with the instantaneous potential values $\phi^i(t)$. The amount of property i that disappears with the reactant molecules is

$$\sigma^i(\text{reactants}) = A_k^i[\phi^i(t)]\sigma_k(\text{reactants}). \tag{79}$$

The amount of property i that appears with the products is

$$\sigma^i(\text{products}) = A_k^i[\phi^i(t)]\sigma_k(\text{products}). \tag{80}$$

In (80) and (79) the functional dependence of the scale factors upon the potentials is purposely included to stress that the scale factors for reactants and products are evaluated at the same potential values: those pertaining at the instant of time t. The reactions are evaluated isopotentially.

It is, of course, almost always the case that the potentials will change during the course of a chemical reaction, provided that pains are not taken to maintain a constant environment by controlling the flow of properties. Consider, for example, the reaction $A \leftrightarrows B + C$, which is occurring at the rate $d\xi/dt$. Suppose $\bar{V}_A[p(t), T(t)] < \bar{V}_B[p(t), T(t)] + \bar{V}_C[p(t), T(t)]$; That is, the partial molar volumes of the products are greater than that of the reactant. The local pressure will go up, or it may be relieved by a divergence in volume flow. Likewise, there may be a change in thermal energy if $\bar{S}_A \neq \bar{S}_B + \bar{S}_C$, the rate of change being $(\bar{S}_B + \bar{S}_C - \bar{S}_A)T\, d\xi/dt$. The change in temperature caused by this change in thermal energy is calculated by the caloric equation, which depends upon constitutive parameters such as the coefficient of heat conduction and the specific heat capacity of the continuum (see Section VIII). At the next moment in time $t + \Delta t$ the scale factors have to be evaluated at the prevailing potential values; for example, $\bar{V}_{\mathrm{A}}[p(t + \Delta t), T(t + \Delta t)]$ or $\bar{S}_{\mathrm{C}}[p(t + \Delta t), T(t + \Delta t)]$.

Above and beyond the changes in thermal energy accounted for by an (isopotential) examination of scale factors, the reaction itself may involve thermal energy. It may be exothermal or endothermal. This energy must be included in any overall energy accounting, and to preserve the formalism it must be included in the s-space representation of thermal energy sT. This is not immediate, and the calculation will occupy the last fourth of this section. At this point it suffices to state that the energy released in a chemical reaction that is not accounted for by a consideration of the scale factors of reactants and products is random scalar energy and hence thermal. The rate of appearance (or disappearance) of this energy is

$$\text{reaction energy rate} = u_R \, d\xi_R/dt, \tag{81}$$

where u_R is the amount of energy released per unit reaction R.

G. Scalar Potential Energy and Maxwell's Relationships

As a prelude to the derivation of the entropy source term by means of the first law of thermodynamics (in effect, the growth equation for scalar potential energy), a few comments based upon the representation of scalar energy should be made. The potential energy per mole of species k is given as a mapping from the potentials:

$$u_k = A_k^i \phi^i. \tag{82}$$

In an analysis of the continuum, the variable of interest is energy density u. This is simply

$$u = C_k u_k = C_k A_k^i \phi^i, \tag{83}$$

which under the mapping $C_k \to a^i$ gives the following theorem.

THEOREM. There is a congruence between the representation of scalar potential energy in p-space and in s-space. The representation is given by the bilinear invariant form

$$C_k u_k = a^i \phi^i. \tag{84}$$

The bilinear expression

$$u = a^i \phi^i \tag{85}$$

is a homogeneous function of the first order in either a^i or ϕ^i. By Euler's theorem, u can be written as

$$u = (\partial u/\partial \phi^i)\phi^i \tag{86}$$

and also as

$$u = (\partial u/\partial a^i)a^i. \tag{87}$$

Comparing the preceding three equations gives two interesting relationships:

$$\partial u/\partial a^i = \phi^i \qquad \text{and} \qquad \partial u/\partial \phi^i = a^i. \tag{88}$$

If now second-order derivatives are taken, one finds the following theorem.

THEOREM. Properties and potentials satisfy the following symmetry conditions belonging to the general class called Maxwell's relationships:

$$\partial \phi^i/\partial a^j = \partial \phi^j/\partial a^i \qquad \text{and} \qquad \partial a^j/\partial \phi^i = \partial a^i/\partial \phi^j. \tag{89}$$

H. The First Law of Thermodynamics

The scalar potential energy u is not a property and, lying outside of that formal category, has no equation of continuity as such. Nevertheless, it is an accountable quantity. Its rate of growth is given by the first law of thermodynamics, a universal observation valid for all systems. In the continuum it is written

$$\partial u/\partial t = -\nabla \cdot \mathbf{J}_q - u_k \nabla \cdot \mathbf{J}_k + \partial w/\partial t. \tag{90}$$

The divergence terms account for changes in the local potential energy (density) due to heat conduction and the convection of energy with particle fluxes. The last term of (90) is the power (time rate of change of work) being deposited in the unit volume under consideration. There are several contributors to this work rate term, which might better be called the energy source term in as far as (90) looks (almost) like an equation of continuity.

Broken into its component parts, the work rate is

$$\frac{\partial w}{\partial t} = \frac{\partial w(\mathbf{J}_k)}{\partial t} + C_k A^i_k \frac{\partial \phi^i}{\partial t} + u_k \sigma_k + u_R \frac{\partial \xi_R}{\partial t}, \tag{91}$$

(i) where the rate of work done to overcome the frictional and viscous forces (i.e., dissipative forces) along the flux trajectories is

$$\partial w(\mathbf{J}_k)/\partial t = \mathbf{J}_k \cdot \mathbf{F}_k. \tag{92}$$

This is an immediate consequence of defining the net force $\mathbf{F}_k$ to the sum of the driving force and the inertial force. The total force (density) available to overcome the dissipative forces is $C_k \mathbf{F}_k = \nabla \cdot \mathbf{T}$ (see Section III.C). The four-vector formulation for $\mathbf{F}_k$ allows the work rate (92) arising from motion to be expressed in a manner free of the constitutive parameters associated with the stress tensor.

(ii) where the input of power to the unit volume due to changes $\partial\phi^i/\partial t$ in the potentials is

$$C_k A_k^i \partial\phi^i/\partial t. \tag{93}$$

The changes $\partial\phi^i/\partial t$ may arise locally, for example, a pressure increase as the consequence of a chemical reaction. Or they may be local manifestations of changes occurring elsewhere; for example, power "pumped" in by varying an external electric field.

(iii) where the source of potential energy from creation and annihilation processes, calculated isopotentially, is

$$u_k \sigma_k \tag{94}$$

(iv) where the source of thermal energy above and beyond that accounted for in (94) is

$$u_R \partial\xi_R/\partial t. \tag{95}$$

It is by way of an examination of the first law of thermodynamics that an analytical expression for the entropy source term σ is discovered. The first law is a statement about physics. In itself it is not sufficient. The calculation evolves from the absolute necessity of a congruence between the representation of energy in s-space and in p-space. The s-space description has been forcibly augmented to include thermal effects not accountable within the formalism as initially postulated; the variables indexed by superscript m were not sufficient. On the preceding pages, the entropy flux was derived. It remains to find σ. This source term will be assigned that functional dependence which is necessary to assure the desired congruence between the representations in s-space and in p-space.

I. Entropy Production and the Dissipation Function

The first law (90)–(95) is a statement about physics given in p-space, with certain terms as required by observations about heat conduction and chemical reactions. How does $\partial u/\partial t$ appear in s-space? From $u = a^i\phi^i$, the answer is immediate:

$$\frac{\partial u}{\partial t} = a^i \frac{\partial\phi^i}{\partial t} + \phi^i \frac{\partial a^i}{\partial t}. \tag{96}$$

The equations of continuity for all properties, excluding $a^m = s$, are given by

(54). That for s is given by (70) with (75) as

$$\frac{\partial s}{\partial t} = -\nabla \cdot \left(\mathbf{J}^m + \frac{1}{T} \mathbf{J}_q \right) + \sigma^m + \sigma. \tag{97}$$

With the equations of continuity for all properties, (96) becomes

$$\frac{\partial u}{\partial t} = a^i \frac{\partial \phi^i}{\partial t} - \phi^i \nabla \cdot \mathbf{J}^i - \nabla \cdot \mathbf{J}_q + \frac{1}{T} \mathbf{J}_q \cdot \nabla T + \phi^i \sigma^i + T\sigma, \tag{98}$$

where, as always, the summations include the index m. Note that $\nabla \cdot (\mathbf{J}_q/T)$ in (97) ends up as two terms in (98).

At this point, there are two representations for $\partial u/\partial t$: (90) and (98). In (98) there is the term $T\sigma$, containing the as yet undefined variable σ. Within the overall congruence established by setting the two equations for $\partial u/\partial t$ equal to each other, there are two congruences that were noted earlier, namely, $u_k \sigma_k = \phi^i \sigma^i$ and $u_k \nabla \cdot \mathbf{J}_k = \phi^i \nabla \cdot \mathbf{J}^i$, by theorems (57) and (62). Thus by the necessary equality of (90) and (98), the entropy source term must be given by

$$T\sigma = -\frac{1}{T} \mathbf{J}_q \cdot \nabla T + \mathbf{J}_k \cdot \mathbf{F}_k + u_R \frac{\partial \xi_R}{\partial t}. \tag{99}$$

The last term of (99) may be positive or negative, depending on whether the creation and annihilation processes absorb or release energy as various bonds of aggregation are broken and others formed. A large negative value for this reaction energy term may result in a negative value for the entropy source term σ. However, the very nature of physical processes imposes a strong condition upon the overall entropy source term $\sigma^s = \sigma^m + \sigma$, namely, $\sigma^s \geq 0$. This is the second law of thermodynamics in its most succinct form.

The first two terms of (99) are the measure of potential energies degraded during motion into thermal energy. They are the measure of the dissipative processes in complex real systems. Together they define the dissipation function

$$\delta = -(1/T)\mathbf{J}_q \cdot \nabla T + \mathbf{J}_k \cdot \mathbf{F}_k. \tag{100}$$

The dissipation function is defined above by a vector equation. On the other hand, the contribution to σ^s from creation and annihilation processes is given by scalar terms. By using a principle attributed to Curie, it can be argued (see, e.g., deGroot and Mazur, 1963) that in isotropic media the scalar and vector contributions to σ^s are not coupled, each being positive definite. In that case, the second law states that the degradation of energy by transport is a one-way process; in short,

$$\delta \geq 0. \tag{101}$$

J. Steady-State and Transient Dissipation

The force $\mathbf{F}_k$ that appears in the dissipation function (100) is the net force. It includes the driving forces and inertial forces. By (6)

$$\mathbf{F}_k = \mathbf{F}_k(u) + \mathbf{F}_k(\mathbf{U}) = -[\mathbf{\nabla}; D_{t,k}][u_k; \mathbf{U}_k], \tag{102}$$

where u_k is the scalar potential energy and $\mathbf{U}_k$ is the vector potential energy, or momentum. Since the net force is the sum of two terms, two special cases immediately present themselves for comment: (1) $F_k(\mathbf{U}) = 0$ and (2) $F_k(u) = 0$.

The situation in which all inertial forces are zero is called the steady state since it springs from a null condition imposed upon the substantial time deviative, namely, $D_{t,k}\mathbf{U}_k = 0$. An example of such a steady-state system is a steel sphere dropping down a vertical column of viscous fluid. At some point it will proceed with a steady velocity. From that point on, the gravitational force on the sphere is exactly balanced by viscous forces. There is no change in momentum along the trajectory. All gravitational potential energy is dissipated to thermal energy.

The situation in which all driving forces are zero is called the transient case because the increase in the dissipation function is achieved by a corresponding decay in momentum. An example of such a transient system is a block of wood sliding upon a horizontal table. Its initial momentum is transformed to thermal energy along the trajectory, with no change being made nonthermal potential energy.

K. Purely Thermal Systems

The distinction in s-space between real-potentials and pseudo-potentials leads to an interesting restriction of the general thermodynamics developed in this chapter. Consider a system in which there is a complete interconvertability between the scalar and the vector parts of any real-potential. That is, the system is ideal insofar as the net fields associated with real-potentials are zero. By the definition (9) of net fields, this condition is, for $\mathbf{\Phi}^i \neq 0$,

$$\mathbf{\Xi}^i = -[\mathbf{\nabla}; D_t][\phi^i; \mathbf{\Phi}^i] = 0. \tag{103}$$

The fields associated with pseudo-potentials are assumed to be nonzero; that is,

$$\Xi^i = -[\mathbf{\nabla}; D_t][\phi^i; 0] \neq 0. \tag{104}$$

It will be best to continue this discussion in more concrete terms. Assume the system can be described by the five most common potentials: gravitational, electrical, informational, thermal, and pressure. Let the net fields associated with the real-potentials of gravity and electricity be zero. This means that any change in gravitational potential energy will be compensated by an equivalent change in momentum, with no energy going into thermal energy. A similar compensation between the electric scalar and vector potentials is implied. No conditions are put upon the pseudo-potentials T, p, and μ_k.

The fields are mapped to forces by the scale factors. In this case

$$\mathbf{F}_k = A_k^i \Xi^i = -\bar{S}_k \nabla T - \bar{V}_k \nabla p - \nabla \mu_k \tag{105}$$

using the scale factors given in Table I. The dissipation function (100) is in this case

$$\delta = -\frac{1}{T} \mathbf{J}_q \cdot \nabla T - \mathbf{J}_k \cdot \bar{S}_k \nabla T - \mathbf{J}_k \cdot \bar{V}_k \nabla p - \mathbf{J}_k \cdot \nabla \mu_k. \tag{106}$$

The scale factors are scalars and can be moved in front of the dots in the scalar products indicated above. The scale factors map particle fluxes to property fluxes, and in particular

$$\mathbf{J}^m = \bar{S}_k \mathbf{J}_k \tag{107}$$

and

$$\mathbf{J}_V = \bar{V}_k \mathbf{J}_k, \tag{108}$$

where $\mathbf{J}_V$ is the flux of volume and $\mathbf{J}^m$ is the convected entropy flux as defined by (71).

The dissipation function (106) can now be written

$$\delta = -[(1/T)\mathbf{J}_q + \mathbf{J}^m] \cdot \nabla T - \mathbf{J}_V \cdot \nabla p - \mathbf{J}_k \cdot \nabla \mu_k. \tag{109}$$

In terms of the entropy flux $\mathbf{J}_s$, defined in (75), this is

$$\delta = -\mathbf{J}_s \cdot \nabla T - \mathbf{J}_V \cdot \nabla p - \mathbf{J}_k \cdot \nabla \mu_k. \tag{110}$$

A system that can be thus described solely in terms of pseudo-potentials is called a strictly thermal system. The dissipation function is a representation in s-space—a sum of fields times their conjugate property fluxes. Entropy flux is conjugate to the temperature field. Volume flux is conjugate to the pressure field. Particle flux is conjugate to the informational field. This tacitly assumes that there is sufficient resolution at the systemic level to identify particle species according to the categories established at the particulate level of description.

VI. VECTOR POTENTIAL ENERGY

A. Conservation Laws

Conservation laws are the most powerful statements in science. They are not, however, primary and certainly not to be taken as axiomatic. In modern theories of the foundations of physics, the conservation laws of linear and angular momentum and of energy are considered to be a direct consequence of the elementary symmetries of nature. They follow from the assertion that the laws of physics are invariant (1) under displacements in time and space, (2) under rotations in space, and (3) under uniform motions. The transformation group formed by these geometrical principles of invariance is the well-known Poincaré group and can be used as an axiomatic basis for the development of special relativity.

There is a second class of conservation laws, concerning nonkinetic properties such as mass and electric charge. Though universally accepted, their basis remains problematic. Wigner (1967) points out that "... while the conservation laws for all other quantities, such as energy or angular momentum, follow in a natural way from the principles of invariance, the conservation law for electric charge so far has defied all attempts to place it on an equally general basis."

By using the formalism developed for complex real systems, it will be demonstrated that the conservation of certain properties is a necessary consequence of the existence of inertial forces and of a dynamical equation for the growth of vector potential energy—and of the necessary congruence of their representations in p-space and s-space. The familiar name for vector potential energy is momentum, and these terms are used interchangeably here. However, it must be kept in mind that this momentum is not restricted to the gravitational (mass) vector potential. It includes all other potentials pertaining to inertial effects, such as the electrical (charge) vector potential. The conservation of mass and all other properties associated with inertial forces (i.e., properties conjugate to real potentials) is thus postulated to be a consequence of the dynamics of momentum. Properties are determined by scale factors. Particles carry properties, and the amount carried is given the mapping $a^i = A^i_k C_k$. The characteristics of the scale factors cannot be postulated *a priori*. They must be derived from dynamic and energetic principles and from requirements related to the congruence of representations. Such considerations, for example, led to Theorem (52), which places a constraint on any possible dependence of the scale factors upon the scalar potentials.

B. Dynamics and Momentum Growth

The sum of the driving force $\mathbf{F}_k(u)$, derived from scalar potentials, and the inertial force $\mathbf{F}_k(\mathbf{U})$, derived from vector potentials, is the net force available to overcome dissipative forces. In four-vector notation the net force is

$$\mathbf{F}_k = -[\mathbf{\nabla}; D_{t,k}][u_k; \mathbf{U}_k]. \tag{111}$$

In continuum mechanics the dissipative force per unit volume is given in terms of the stress tensor as $-\mathbf{\nabla} \cdot \mathbf{T}$. The equality of net force and dissipative force is thus

$$C_k \mathbf{F}_k = \nabla \cdot \mathbf{T}, \tag{112}$$

taking into account that they act in opposite directions. This equation defines the dynamics and is simply a statement of dynamical balance: the total net force available for opposing dissipative forces is equal to the dissipative force actually present in the continuum.

There is a second (and independent) fundamental equation relating vector potentials to forces and stresses that is analogous to the relationship provided for scalar potentials by the first law of thermodynamics. The first law is, in essence, a growth equation for scalar potential energy. The growth of vector potential energy (momentum) is described by a similar balance equation, which ranks among the first and second laws of thermodynamics as a basic principle governing the behavior of complex real systems. It is given by a generalization of the Cauchy equation of hydrodynamics and accounts for the growth in total moment $\mathbf{U}$, not just mass momentum:

$$\partial \mathbf{U}/\partial t = -\nabla \cdot \mathbf{U}_k \mathbf{J}_k - \nabla \cdot \mathbf{T} + C_k \mathbf{F}_k(u). \tag{113}$$

The first term on the right-hand side represents momentum losses to the unit volume by convection. The second term represents losses to dissipative forces. The last term represents the gain of momentum arising from the action of driving forces. This term is analogous to the familiar increase of kinetic energy that compensates a loss of potential energy in analytical mechanics (the conservation of energy in ideal systems of particles). Before attention is directed to the analysis of scale factors and the derivation of conservation laws, a short discussion of the role of momentum convection and stresses in the growth equation (113) should be presented. An equation that lies at the core of dynamics warrants elaboration.

C. Convection of Momentum

The total momentum (density) is the sum of the momenta associated with the various particles species; that is,

$$\mathbf{U} = C_k \mathbf{U}_k, \tag{114}$$

where the species momentum is given by the mapping $\mathbf{U}_k = A^i_k \mathbf{\Phi}^i$. Finding the rate of increase of momentum due to convection is similar to finding the rate of increase of a property, $\partial a^i/\partial t$. The flux of particles is $\mathbf{J}_k = C_k \mathbf{v}_k$, and since each mole of particle carries an amount of property A^i_k, the property flux is $\mathbf{J}^i = A^i_k \mathbf{J}_k$. The rate of increase (per unit volume) is simply

$$\partial a^i/\partial t = -\nabla \cdot A^i_k \mathbf{J}_k, \tag{115}$$

assuming only convection and no reactions. In the case of momentum convection, each mole of particle k carries the momentum $\mathbf{U}_k$. Properties and scale factors are scalars, and the calculation of a flux of property was immediate. Obviously, it is not possible to transform a vector quantity such as momentum into a convective vector flux. The calculation requires a "flux" of one tensorial order higher—a dyadic.

In Fig. 2 is shown one side of a volume, of dimension Δx, Δy, Δz. Consider the x-component U_{kx} of the momentum $\mathbf{U}_k$ associated with species k. In Fig. 2 the vertical side at x has an area $\Delta y\ \Delta z$. The particle flow through it and into the volume is $J_{kx}(x)\,\Delta y\,\Delta z$. The particle flow $\mathbf{J}_k$ has vector components (J_{kx}, J_{ky}, J_{kz}). Convected with this particle flow is a flow of the x-component of momentum equal to $U_{kx}(x)J_{kx}(x)\,\Delta y\,\Delta z$. Through the opposite side at $x + \Delta x$, the particle flow out of the volume element is $J_{kx}(x + \Delta x)\,\Delta y\,\Delta z$. The outward flow of the x-component of momentum is thus $U_{kx}(x + \Delta x)\,J_{kx}(x + \Delta x)\,\Delta y\,\Delta z$. This cross flow results in a time change per unit volume of

$$\frac{\partial U_{kx}(\text{cross})}{\partial t} = \lim \frac{[U_{kx}(x)J_{kx}(x) - U_{kx}(x + \Delta x)J_{kx}(x + \Delta x)]\,\Delta y\,\Delta z}{\Delta x\,\Delta y\,\Delta z}$$

$$= -\frac{\partial}{\partial x}(U_{kx}J_{kx}). \tag{116}$$

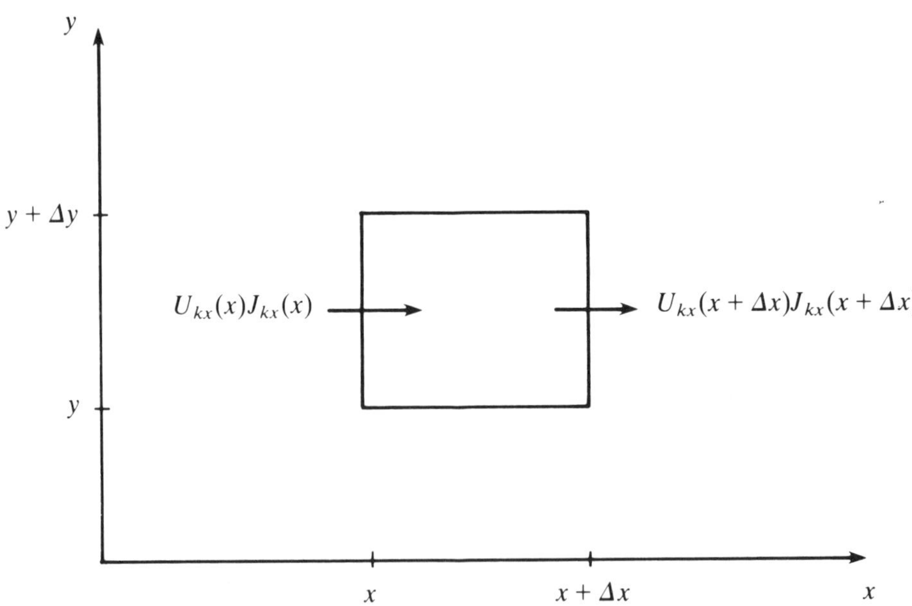

FIG. 2.

Similarly, in the y-direction,

$$\frac{\partial U_{kx}(\mathrm{up})}{\partial t} = -\frac{\partial}{\partial y}(U_{kx}J_{ky}). \tag{117}$$

and in the z-direction,

$$\frac{\partial U_{kx}(\mathrm{out})}{\partial t} = -\frac{\partial}{\partial z}(U_{kx}J_{kz}). \tag{118}$$

The sum of these three contributions gives the net change in the x-component of $\mathbf{U}_k$ that results from convection:

$$\frac{\partial U_{kx}}{\partial t} = -\boldsymbol{\nabla} \cdot U_{kx}\mathbf{J}_k. \tag{119}$$

The same procedure gives the change in the y-component of $\mathbf{U}_k$ as

$$\frac{\partial U_{ky}}{\partial t} = -\boldsymbol{\nabla} \cdot U_{ky}\mathbf{J}_k \tag{120}$$

with a similar expression for $\partial U_{kz}/\partial t$. Thus it is seen that the total change in the momentum vector $\mathbf{U}_k$ resulting from convection is

$$\frac{\partial \mathbf{U}_k}{\partial t} = -\boldsymbol{\nabla} \cdot \mathbf{U}_k\mathbf{J}_k, \tag{121}$$

where $\mathbf{U}_k\mathbf{J}_k$ is a dyad. After summing the right-hand side of (121) over k, one obtains the total momentum change for all species—this is the term that appears in the momentum growth equation (113).

D. The Stress Tensor

Momentum changes resulting from the presence of dissipative (viscous) forces are accounted for by means of the stress tensor $\mathbf{T}$. The dashed lines in Fig. 3 enclose two volumes of fluid flowing adjacent to each other. Notice that they are flowing with different velocities along the x-axis. Assume that v_y and v_z are constant.

For the sake of the argument, let $v_x(y) > v_x(y + \Delta y)$. The fluid is not a rigid body. There is slippage between the two volume elements. With slippage, there is a drag force, that is, a viscous force. The faster volume element with velocity $v_x(y)$ transfers some of its momentum to the slower volume element with velocity $v_x(y + \Delta y)$. In other words, some x-component of momentum is transferred in the y-direction. The amount transferred (per unit time and unit area) is denoted T_{yx}. In general, when v_y and v_z are not constant, there are momentum transfers associated with all other directions, that is, with all

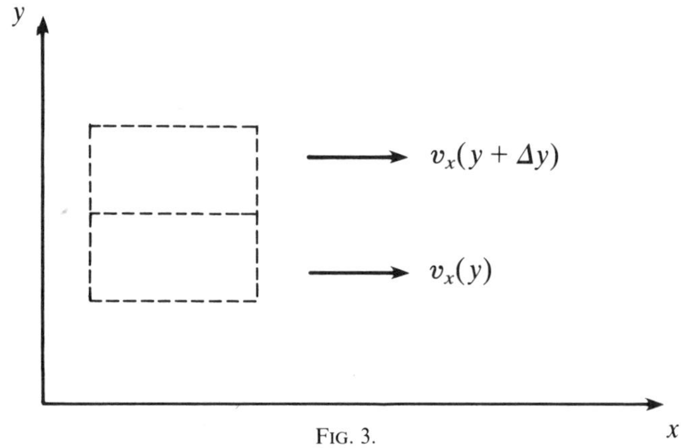

FIG. 3.

binary permutations of (x, y, z). These elements comprise the nine components of the stress tensor **T**.

For simple, homogeneous fluids, Newton's law of viscosity assumes that the momentum transfer is proportional to the velocity gradient; thus for the situation depicted in Fig. 3

$$T_{yx} = -\eta\, \partial v_x/\partial y, \tag{122}$$

where η is the viscosity coefficient. In the strict sense of the word, this is not a law; it is a model. There are many other models for the stress tensor, some of remarkable complexity. In the analysis of the dynamics of complex real systems, these models are of little concern. What matters is the general formulation of momentum transfer by viscous forces. In terms of the hierarchy of descriptions, viscous forces are resolved only at a level lower than the level chosen for the p-space description. Forces resolvable at the p-space level are accounted for explicitly in the momentum growth equation (113) by the $C_k\mathbf{F}_k(u)$ term. The stress tensor is so defined that the momentum transfer is to a unit volume, not specifically to particles. It is a representation in s-space; and, just as pressure and temperature, it is a manifestation at the s-space level of description of interactions that cannot be resolved in the chosen p-space. The stress tensor is not a potential, however. The pseudo-potentials p and T provide an accounting for degraded energies, whereas the stress tensor characterizes the interactions by which these energies were degraded.

The calculation of the contribution of momentum transfer by viscous forces starts with a consideration of Fig. 2. The changes in the volume element of each component of **U** are considered separately. What is $\partial U_x/\partial t$? The stress tensor has units of rate of momentum change per unit area. Hence, the change in U_x due to momentum transfer through the vertical side at x is $T_{xx}(x)\,\Delta y\,\Delta z$. The change through the vertical side at $x + \Delta x$ is $T_{xx}(x + \Delta x)\,\Delta y\,\Delta z$. This gives

$$\partial U_x(\text{across})/\partial t = -\partial T_{xx}/\partial x. \tag{123}$$

The change in U_x through the sides of the volume element that are perpendicular to the y-axis is

$$\partial U_x(\text{up})/\partial t = -\partial T_{yx}/\partial y \tag{124}$$

and the change through the sides perpendicular to the z-axis is

$$\partial U_x(\text{out})/\partial t = -\partial T_{zx}/\partial z. \tag{125}$$

Similar calculations for the y- and z-components of momentum lead to a total change due to transfer by viscous forces of

$$\partial \mathbf{U}/\partial t = -\boldsymbol{\nabla}\cdot\mathbf{T}. \tag{126}$$

E. Conservation of Properties

At this point, two separate and independent dynamical equations have been presented. Equation (112) defines the net force available to overcome dissipative forces. Equation (113) is a statement accounting for the growth of momentum (vector potential energy). They must be compatible. In fact, a law for the conservation of properties associated with real-potentials is a necessary consequence of this compatibility. The divergence of the stress tensor is common to both equations. This term, which depends upon constitutive parameters of the continuum, can be eliminated by subtraction of (112) from (113) to give the more general equation

$$\partial \mathbf{U}/\partial t + \boldsymbol{\nabla}\cdot\mathbf{U}_k\mathbf{J}_k = -C_k\mathbf{F}_k(\mathbf{U}). \tag{127}$$

With the definition of inertial force and under the mapping of this p-space representation to s-space, the preceding equation becomes

$$(\partial/\partial t)C_k A_k^i \boldsymbol{\Phi}^i + \boldsymbol{\nabla}\cdot C_k A_k^i \boldsymbol{\Phi}^i \mathbf{v}_k = C_k A_k^i D_{t,k}\boldsymbol{\Phi}^i. \tag{128}$$

Implicit in this transformation is the postulate that A_k^i and $D_{t,k}$ commute. This condition assumes the congruence between the p-space and the s-space representations of inertial force. This commutativity condition is diagramed in Fig. 4. This commutativity diagram for vector potentials is the analogue of the commutativity diagram for scalar potentials in Fig. 1.

A summation over k allows the first term on the left-hand side of (118) to be written as

$$\frac{\partial}{\partial t}(a^i\boldsymbol{\Phi}^i) = a^i\frac{\partial \boldsymbol{\Phi}^i}{\partial t} + \boldsymbol{\Phi}^i\frac{\partial a^i}{\partial t}. \tag{129}$$

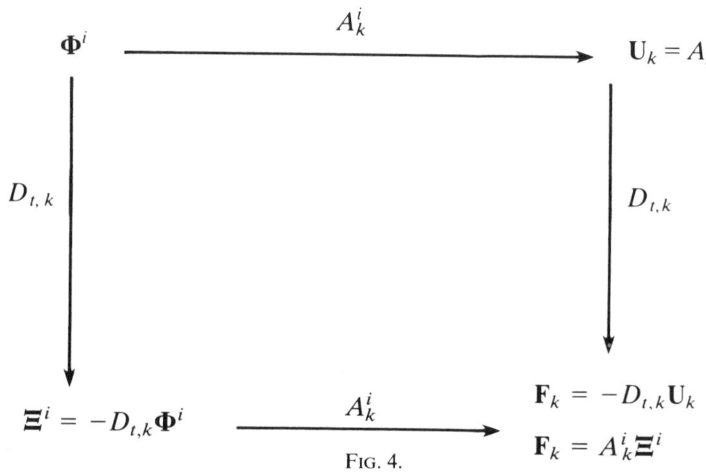

FIG. 4.

The second term on the left-hand side of (128) is a dyadic, which can be written as

$$\nabla \cdot C_k A_k^i \mathbf{\Phi}^i \mathbf{v}_k = C_k A_k^i \mathbf{v}_k \cdot \nabla \mathbf{\Phi}^i + \mathbf{\Phi}^i \nabla \cdot C_k A_k^i \mathbf{v}_k. \tag{130}$$

The substantial derivative on the right-hand side of (118) can be expanded to

$$C_k A_k^i [\partial \mathbf{\Phi}^i/\partial t + (\mathbf{v}_k \cdot \nabla)\mathbf{\Phi}^i] = a^i\, \partial \mathbf{\Phi}^i/\partial t + C_k A_k^i \mathbf{v}_k \cdot \nabla \mathbf{\Phi}^i \tag{131}$$

using (29). This term-by-term examination of the growth equation (128) results in a significant reduction. Equation (127) can now be written

$$\mathbf{\Phi}^i\, \partial a^i/\partial t = -\mathbf{\Phi}^i \nabla \cdot C_k A_k^i \mathbf{v}_k. \tag{132}$$

This equation can be reduced even further since $C_k A_k^i \mathbf{v}_k = \mathbf{J}^i$; thus

$$\mathbf{\Phi}^i(\partial a^i/\partial t + \nabla \cdot \mathbf{J}^i) = 0. \tag{133}$$

The result of the preceding calculation is, then, a linear equation for the vector potentials $\mathbf{\Phi}^i$. However, the vector potentials are, by their physical basis, linearly independent. Therefore, the coefficients in (133) must necessarily be identically equal to zero. This gives the following theorem.

THEOREM. The equation of continuity for any property a^i associated with (conjugate to) a real-potential $[\phi^i; \mathbf{\Phi}^i]$ is

$$\partial a^i/\partial t = -\nabla \cdot \mathbf{J}^i. \tag{134}$$

COROLLARY. Properties conjugate to real-potentials are conserved. That is, the source terms are equal to zero: $\sigma^i = 0$.

F. Consequences of the Conservation of Properties

The equation of continuity for particle concentration is

$$\partial C_k/\partial t = -\nabla \cdot \mathbf{J}_k + \sigma_k. \tag{135}$$

The congruence between the accounting of particles, $\partial C_k/\partial t$, and the accounting of properties $\partial a^i/\partial t$ required by the mapping $a^i = A^i_k C_k$ was discussed in Section IV.D. The condition for congruence is given by (55) to be

$$A^i_k \sigma_k + (\mathbf{J}_k \cdot \nabla \phi^j + C_k\, \partial \phi^j/\partial t)\, \partial A^i_k/\partial \phi^j = \sigma^i. \tag{136}$$

By the preceding theorem, $\sigma^i = 0$ for all values of the index i denoting real-potentials. The scale factors may be functions of the potentials $\{\phi^i\}$, but are not functions of the particle source terms $\{\sigma_k\}$. Therefore, general conclusions about the behavior of $\partial A^i_k/\partial \phi^j$ can be drawn from (136) under the restriction of no reactions: that is, under $\sigma_k = 0$. This gives

$$(\mathbf{J}_k \cdot \nabla \phi^j + C_k\, \partial \phi^j/\partial t)\partial A^i_k/\partial \phi^j = 0. \tag{137}$$

It should be noted that the indices are such that (1) there is a summation over j, (2) the equation holds for any value of k, (3) but it is restricted to those values of i that denote real-potentials. The functions in the parentheses are linearly independent; therefore, the coefficients are zero for all values of j. This holds for any value of k, hence the following theorem.

THEOREM. The scale factors associated with real-potentials are not functions of scalar potentials; that is,

$$\partial A^i_k/\partial \phi^j = 0 \qquad \begin{array}{l}\text{for all } j \text{ and } k \text{ and}\\ \text{for } i \text{ such that } \mathbf{\Phi}^i \neq 0.\end{array} \tag{138}$$

COROLLARY. The scale factors associated with real-potentials are constants.

As mentioned before, the scale factors are independent of the source terms σ_k. This means that (118) is a general condition and can be incorporated into (116) for those values of i corresponding to real-potentials. For real-potentials, $\sigma^i = 0$, and thus (116) gives the following theorem.

THEOREM. The kinetics for creation and annihilation processes are not unconstrained. The source terms must obey the condition

$$A^i_k \sigma_k = 0 \tag{139}$$

for all values of i corresponding to real-potentials.

From another viewpoint, Theorem (139) states that the scale factors of products are related to those of reactants in a manner determined by the reaction kinetics. These scale factors are determined by the mode of aggregation that underlies the reaction. There is a compatibility between scale factors and reaction source terms because they both depend upon how things are put together.

The close relationship between Theorems (138) and (139) can be seen in the physics of the ionization of a salt. Consider a neutral salt molecule C composed of two ions A^+ and B^-. When they are constituents of a salt molecule, ions A^+ and B^- are immersed in a local electrostatic potential $\psi_e(C)$. This potential is a measure of the interaction between A^+ and B^- which generates the force of aggregation holding them together as a single molecule C. Upon ionization, A^+ and B^- find themselves in a different environment, immersed in potentials $\psi_e(A^+)$ and $\psi_e(B^-)$. Theorem (139) states that the charge carried by a particle is independent of the electrostatic potential. Therefore A^+ carries a unit negative charge in isolation as an ion and in concert with B^- as a molecule; in short, $Z_A[\psi_e(A^+)] = Z_A[\psi_e(C)]$. The same holds for B^-. It must be concluded that the charge on C is the sum of the charges on its constituents: that is, $Z_C = Z_A + Z_B = 1 - 1 = 0$. This argument contains Theorem (139) implicitly. The form of the source terms σ_k are implied in the initially stated stoichiometry of the ionization reaction.

It is obvious that the logic of the preceding discussion can be tightened to give the following corollary to Theorem (139).

Corollary. In a hierarchical system where the creation and annihilation of particles occur by aggregation processes, the scale factors associated with real-potentials obey an additive stoichiometry. If reactants are denoted by $k = \mathrm{A}, \mathrm{B}, \ldots$, and products by $k = \mathrm{M}, \mathrm{N}, \ldots$, and if the reaction is governed by the DeDonder equation, $\sigma_k = \nu_k \, d\xi/dt$, then by (139)

$$A^i_{\mathrm{A}}\nu_{\mathrm{A}} + A^i_{\mathrm{B}}\nu_{\mathrm{B}} + \cdots + A^i_{\mathrm{M}}\nu_{\mathrm{M}} + A^i_{\mathrm{N}}\nu_{\mathrm{N}} + \cdots = 0 \tag{140}$$

for all i such that $\Phi^i \neq 0$. The ν_k are the stoichiometric coefficients characteristic of the reaction resulting in the given aggregations.

G. Pseudo-Potentials and Gauge

The commutativity of the operator $D_{t,k}$ and the mapping A^i_k shown in Fig. 4 was central to the argument leading to Theorem (138). This theorem states that the scale factors associated with real-potentials are not functions of scalar potentials. Earlier, in Section IV, the commutativity of the operator ∇ and the mapping A^i_k shown in Fig. 1 was used to establish Theorem (52), restated here.

$$\phi^i \, \partial A^i_k / \partial \phi^j = 0. \tag{141}$$

In light of the fact that $\partial A^i_k/\partial\phi^j = 0$ for vector potentials, the range of the indices i and j in the preceding equation can be restricted to those values denoting pseudo-potentials. Besides putting restrictions upon the functional dependence of scale factors upon potentials, this equation says something important about the potentials themselves. The pseudo-potentials must have fixed origins for their scales. Their values must be prescribed in some absolute system.

In classical dynamics the potentials themselves do not appear in equations, only their derivatives. Suppose a potential ϕ appears always in combination with a linear differential operator D. Consider the transformation $\phi \to \phi + f$, where $Df = 0$. This transformation is said to be gauge invariant since

$$D(\phi + f) = D\phi + Df = D\phi \tag{142}$$

The simplest gauge invariant transformation is obviously $f = \text{const}$.

For example, the four-vector force $\mathbf{F}_k = -A^i_k[\nabla; D_{t,k}][\phi^i; \mathbf{\Phi}^i]$ is invariant under a transformation of the potentials effected by adding a constant. Equation (141) is not. As a necessary consequence, one has the following theorem.

THEOREM. Pseudo-potentials have gauge.

The familiar pseudo-potentials, temperature and pressure, have gauge. Their scales can be altered by changing physical units, but their origins cannot be shifted from zero. As will be seen in Section VIII.C, the informational potential μ_k also has gauge.

H. Perpetual Motion

The universal dictum against perpetual motion of the first kind was used in Section IV to establish that the driving force $\mathbf{F}_k(u)$ must necessarily be the gradient of a scalar function, namely, the scalar potential energy u_k. Since $\mathbf{F}_k(u) = -\nabla u_k$ is a necessary and sufficient condition for the work integral to be zero over any arbitrary closed path in space, the converse holds. That is, if one bases dynamics upon gradient forces, then perpetual motion is prohibited. Once it was established that the dynamics of complex real systems should be based upon gradient driving forces, it was pointed out in Section IV.C that there had to be a congruence between the force representation in p-space and in s-space. The mapping from potential to potential energy had to be congruent with the mapping from field to force. This gave rise to the commutativity diagram in Fig. 1. The condition required for commutativity, as stated in Theorem (52), limits the functional dependence of the scale factors upon scalar potentials.

An analogous commutativity diagram for vector potentials has been used in the present section to derive a series of theorems bearing upon the conservation of properties associated with vector potentials. For example, Theorem (138) states that the scale factors associated with vector potentials are independent of the scalar potentials. Hence it seems only natural to examine the substantial derivative (the gradient in time) and ask if and how it might bear upon the question of perpetual motion.

The work increment in space is $\mathbf{F}_k \cdot d\mathbf{s}$. The analogous "work" increment in time is $\mathbf{F}_k \, dt_k$, where dt_k is the time increment along the trajectory of species k determined by the velocity $\mathbf{v}_k$. Since there is no "closed path" in time as there can be in space, the line integral (42) must be expressed in a different form if it is to serve as an analogue for an integral in time. It is mathematically equivalent to state that the value of the line integral is independent of the path, depending only upon the initial and final points. That is,

$$\oint \mathbf{F}_k \cdot d\mathbf{s} = 0 \Leftrightarrow \int_1^2 \mathbf{F}_k \cdot d\mathbf{s} = g_k(2) - g_k(1), \tag{143}$$

In this form it is immediately evident that $u_k = -g_k$ and $\mathbf{F}_k = -\nabla u_k$. Recall from Section IV that the integral (143) is done statically. Therefore g_k is determined up to an arbitrary function of time.

In the time domain the stricture against perpetual motion is by analogy to (143) simply

$$\int_1^2 \mathbf{F}_k \, dt_k = \mathbf{G}_k(2) - \mathbf{G}_k(1). \tag{144}$$

It is immediately evident that the necessary and sufficient condition for (144) is that $\mathbf{F}_k = D_{t,k}\mathbf{G}_k$. Obviously, this is related to the vector potential energy by a sign change that makes things flow downhill: $\mathbf{U}_k = -\mathbf{G}_k$. The stipulation that things flow downhill is, by the way, just another way of stating the second law—it is the stricture against perpetual motion of the second kind.

The function $\mathbf{G}_k$ is given by considerations of the time integral (144) and is therefore determined up to an arbitrary function of space. The total force per mole of species k is then the following sum:

$$\mathbf{F}_k = -\nabla u_k - D_{t,k}\mathbf{U}_k \equiv -[\nabla; D_{t,k}][u_k; \mathbf{U}_k]. \tag{145}$$

It is concluded that a necessary and sufficient condition that perpetual motion of the first kind is prohibited is that the force be given by the four-vector gradient operating on the four-vector potential energy. In integral form this is stated

$$\int_1^2 \mathbf{F}_k \cdot d[\mathbf{s}; t_k] = -[u_k; \mathbf{U}_k]_1^2. \tag{146}$$

I. Lorentz Forces

The Lorentz formulation of the ponderomotive force of electricity was presented in Section III.F. It was demonstrated that the four-vector gradient operating upon the four-vector potential gave this familiar force. However, the Lorentz force does not appear in any of the subsequent discussion of dynamics. It did not appear in the definition (25) of the system dynamics, nor in the more explicit dynamics of (113). How and why, then, does it arise in physics? It arises because of an asymmetrical rearrangement of the momentum growth equation (113). Instead of an equation treating all momenta on an equal basis, an equation is written that focuses attention upon the growth of mass–momentum.

Those terms of (113) associated with the dynamics of mass are

$$\text{momentum change} = \frac{\partial(\rho_k \mathbf{v}_k)}{\partial t} \tag{147}$$

$$\text{momentum convection} = \nabla \cdot \rho_k \mathbf{v}_k \mathbf{v}_k \tag{148}$$

and

$$\text{driving force} = -C_k \bar{M}_k \nabla \psi_g = -\rho \nabla \psi_g \tag{149}$$

where $\rho_k = C_k \bar{M}_k$ (no summation) is the contribution of species k to the total mass density ρ. Thus (113) can be written

$$\frac{\partial}{\partial t} \rho_k \mathbf{v}_k + \nabla \cdot \rho_k \mathbf{v}_k \mathbf{v}_k + \nabla \cdot \mathbf{T} = -\rho \nabla \psi_g - \frac{\partial}{\partial t} \sum{}^* C_k A_k^i \mathbf{\Phi}^i$$
$$-\sum{}^* \nabla \cdot C_k A_k^i \mathbf{\Phi}^i \mathbf{v}_k + \sum{}^* C_k \mathbf{F}_k(u), \tag{150}$$

where the notation Σ^* means a summation over all potentials $\mathbf{\Phi}^i$, excluding the gravitational potential. The Lorentz force is still not apparent. The expansion (130) and the conservation equation (132) for properties associated with real-potentials allow the second and third terms on the left-hand side of (150) to be written as

$$C_k A_k^i \, \partial \Phi^i / \partial t + D_k A_k^i \mathbf{v}_k \cdot \nabla \mathbf{\Phi}^i = C_k A_k^i D_{t,k} \mathbf{\Phi}^i \tag{151}$$

in terms of the substantial derivative. The substantial derivative was expanded in (32) to include the term $\nabla \times \mathbf{\Phi}^i$. With this, the dynamics (150) becomes

$$\frac{\partial(\dot{\rho}_k \mathbf{v}_k)}{\partial t} + \nabla \cdot \rho_k \mathbf{v}_k \mathbf{v}_k + \nabla \cdot \mathbf{T}$$
$$= -\rho \nabla \psi_g - \sum{}^* a^i \frac{\partial \mathbf{\Phi}^i}{\partial t} + \sum{}^* C_k A_k^i \mathbf{v}_k \times (\nabla \times \mathbf{\Phi}^i)$$
$$-\sum{}^* C_k A_k^i \nabla_{\mathbf{\Phi}} (\mathbf{v}_k \cdot \mathbf{\Phi}^i) - \sum{}^* C_k A_k^i \nabla \mathbf{\Phi}^i. \tag{152}$$

It is seen that there is a Lorentz force term for every vector potential $\mathbf{\Phi}^i$, with the exception of $\mathbf{\Phi}_g = \mathbf{v}_k$. Note that the last term of (152) includes the driving forces of all scalar potentials (real and pseudo), with the exception of $\psi_g = \phi_g$. The preceding calculation of the Lorentz force is, in the end, simply a combination of the two fundamental and independent formulations of the dynamics, (112) and (113), augmented by the expansion for $D_{t,k}$ given by (32)

VII. FIELD EQUATIONS

A. Potential Functions

The motion of a particle is an observable manifestation of its interaction with all other particles in the cosmos. It is, to be more precise, a manifestation of an asymmetry in the sum total of all interactions. Force is calculated as the four-vector gradient of the four-vector potential, and so a finite force represents a spatial or temporal nonuniformity in the potentials. Thus the science of dynamics and energetics ultimately reduces to the formulation of the equations that characterize the interactions between particles—the field equations for the real-potentials.

As long ago as 1845, Gauss perceived that a physics of the motions of charged bodies could be based upon the consideration of interactions propagated at a finite velocity. As was his wont with research that had not led to a complete solution, he did not publish the results of these studies. In a posthumous note, his pupil Riemann demonstrated that an interaction described by a retarded integral is mathematically equivalent to a propagated action described by the wave equation. At the turn of the century the theory of electricity based upon retarded potentials was essentially completed with the work of Lorentz and Poincaré.

B. Scalar Retarded Potential

Conceptually, there are few definitions in physics simpler than that for a retarded potential. The intensity of the static interaction of a unit test particle at the point P in Fig. 5 with all other particles in the cosmos is expressed quantitatively by a scalar potential function ϕ. This potential should carry an index i to indicate that the interaction of interest is related to the property a^i. One writes ϕ^i. This intensity of interaction is at point P at time t. One writes $\phi^i(P, t)$. It is significant that P is not written in boldface; it denotes a location but not its radius vector in some general coordinate system.

The potential $\phi^i(P, t)$ is proportional to the amount of property i situated in the volume dV' at the source point S'. The whole idea of retarded potentials is

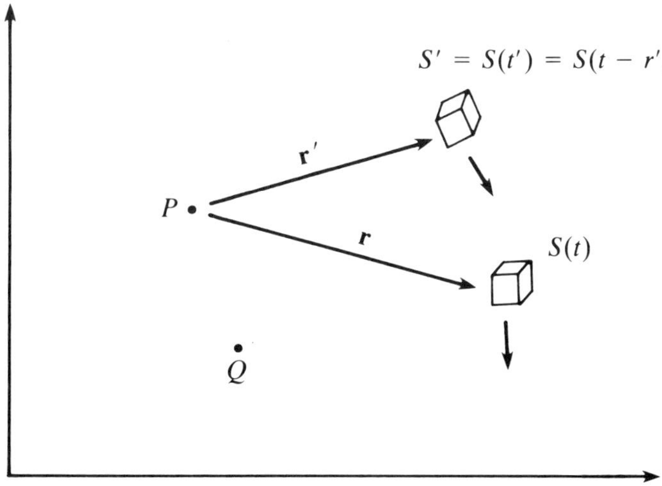

FIG. 5.

contained in the primes used in the preceding sentence. At time t an action emitted from a source element is received at the point P. At that instant the source element is at the point S, or, to be quite explicit, $S(t)$. At that instant the radius vector to the source element is $\mathbf{r}$. This radius vector is in a local coordinate system with the point P as the origin. It measures positions and distances relative to the point P. However, in accord with the basic premise that actions travel with a finite velocity, the vector $\mathbf{r}$ to the source element at the point $S(t)$ does not give the position of the source at the moment when the action was emitted. The action was emitted at some earlier time t', when the source was at the point S'. At the moment of emission, the source was at the position given by $\mathbf{r}'$.

The determination of the retarded time t' is simple. The action travels at the constant velocity c. This action travels the distance r' between the emitter at S' and the receiver at P. Hence the retarded time is

$$t' = t - r'/c. \tag{153}$$

With this determination of the time and position of the source term at the moment of the emission of the action received at P, it is possible to calculate $\phi^i(P, t)$. The potential, as the measure of interaction, depends upon how much property i was at S' at t'. That is, the strength of an interaction is proportional to the amount of property giving rise to the interaction. The density of property i at S' is $a^i(\mathbf{r}', t')$, and the volume of the source element is dV'; thus

$$\phi^i(P, t) \propto a^i(\mathbf{r}', t - r'/c)\, dV'. \tag{154}$$

It is assumed that the strength of the interaction diminishes with distance from the source. It is assumed that it falls off as the inverse of the relative distance r' between emitter and receiver; thus

$$\phi^i(P,t) \propto 1/r'. \tag{155}$$

Combining the effects of source strength and of distance in an integral over all space gives the final form for the scalar potential function as

$$\phi^i(P,t) = K^i \int \frac{a^i(\mathbf{r}', t - r'/c)}{r'} dV', \tag{156}$$

where K^i is a normalization constant. This is an empirically determined constant characteristic of a given interaction and is put into units that give a common scaling of force for all values of i.

C. Vector Retarded Potential

The scalar potential depended upon the static configuration in space of property i at the moment that an action was emitted from the source point S'. The scalar potential was thus determined by the relative distance r'. The vector potential depends upon the relative motion of the receiver at point P and the emitter at point S'. It is thus determined by the relative velocity vector $\dot{\mathbf{r}}'$. To be precise, the vector potential is determined by the relative motions of property i; that is,

$$\mathbf{\Phi}^i(P,t) \propto \mathbf{J}^i(\mathbf{r}', t - r'/c), \tag{157}$$

where, as before, the primes refer to the retarded position S'. The property flux $\mathbf{J}^i$ is given relative to the coordinates centered at P; explicitly,

$$\mathbf{J}^i(\mathbf{r}', t - r'/c) = A^i_k C_k(\mathbf{r}', t - r'/c)\dot{\mathbf{r}}'_k, \tag{158}$$

where $\dot{\mathbf{r}}'_k$ is the velocity of species k relative to the point P.

As before, it is assumed that the vector potential falls off inversely with the distance r'. Hence the vector retarded potential is given by the following integral over all space:

$$\mathbf{\Phi}^i(P,t) = \frac{K^i}{c^2} \int \frac{\mathbf{J}^i(\mathbf{r}', t - r'/c)}{r'} dV'. \tag{159}$$

The normalization to a common scale for forces by means of the constant K^i is the same for scalar and vector potentials. The further normalization of (159) by $1/c^2$ arises because of the mismatch of units between the space and time operators, $\mathbf{\nabla}$ and D_t, and between the space and time potentials, ϕ^i and $\mathbf{\Phi}^i$. This normalization results in an agreement in units in the realm of forces and fields: that is, $\mathbf{\nabla}\phi^i$ has the same units as $D_t\mathbf{\Phi}^i$.

This unsymmetric normalization of the scalar and vector potentials is disquieting and calls for comment. In the metric goemetry of relativity theory, space and time are brought to common units by a change in the scale of time. Space and time are unified in a common metric if physics is framed in a time system measured by $\tau \equiv ct$. In this system, the two potentials are normalized by the same constant K^i. This is expressed by the notation, $\bar{\phi}^i(K^i)$ and $\bar{\mathbf{\Phi}}^i(K^i)$, where the bar over the variables denotes the τ-time system. The fields are given by $\Xi^i = [\mathbf{V}; D_\tau][\bar{\phi}^i(K^i); \bar{\mathbf{\Phi}}^i(K^i)]$. In this system, $\mathbf{V}$ and D_τ have the same units and $\bar{\phi}^i$ and $\bar{\mathbf{\Phi}}^i$ have the same units. Now, with the physics of fields properly formulated in space–time, how can it be transformed back to ordinary t-time? It is simple. The operator $D_\tau = (1/c)D_t$ and since $\bar{\mathbf{J}}^i = (1/c)\mathbf{J}^i$, the vector potential transforms as $\bar{\mathbf{\Phi}}^i(K^i) = (1/c)\mathbf{\Phi}^i(K^i)$. This field, Ξ^i, must be invariant under coordinate transformations. This means that $[\mathbf{V}; D_\tau][\bar{\phi}^i(K^i); \bar{\mathbf{\Phi}}^i(K^i)] = [\mathbf{V}; (1/c)D_t][\phi^i(K^i); (1/c)\mathbf{\Phi}^i(K^i)] = [\mathbf{V}; D_t][\phi^i(K^i); \mathbf{\Phi}^i(K^i/c^2)]$, since the scalar potential is independent of the change in time scale. Aside from making one wonder where the $1/c^2$ came from, the familiar normalization, K^i/c^2, of classical electrodynamics causes no problems.

The terms *potential* and *retarded potential* have been used interchangeably. There is no difference. In the analysis of real complex systems, all real-potentials are retarded potentials.

D. Imbedding the Potentials in a Common Coordinate System

The retarded potentials have been formulated in terms of distances and velocities relative to the given point, P. The same procedure is followed to find the retarded potentials at another point, Q. Values of the potentials at the two points can be compared. It may be that the scalar potential at P is greater than that at Q. However, it is impossible at this moment to calculate the resulting force experienced at P because of this nonuniformity in potential. There is no way to calculate this gradient of the potential given $\phi^i(P, t)$ and $\phi^i(Q, t)$ as the only information. The points P and Q need to have their locations specified in a common coordinate system. This is simple enough. Figure 6 shows the points P and Q situated in a common system. The position of P is $\mathbf{R}(P)$. The retarded distance to the source is $\mathbf{r}' = \mathbf{R}(S') - \mathbf{R}(P)$. The scalar retarded potential (156) can be integrated using $\mathbf{r}'$-coordinates, or, under a simple transformation, using $\mathbf{R}$-coordinates. Thus the notation for the scalar potential evaluated at an arbitrary point P, having the position $\mathbf{R}$, is $\phi^i(\mathbf{R}, t)$.

Knowing the position $\mathbf{R}$ of the point P is not sufficient to embed the vector retarded potential (159) in a common coordinate system. The potential depends upon the flux of property i at the source point S'. This flux is determined relative to the point P; it is proportional to $\dot{\mathbf{r}}'$, as shown in (158). As

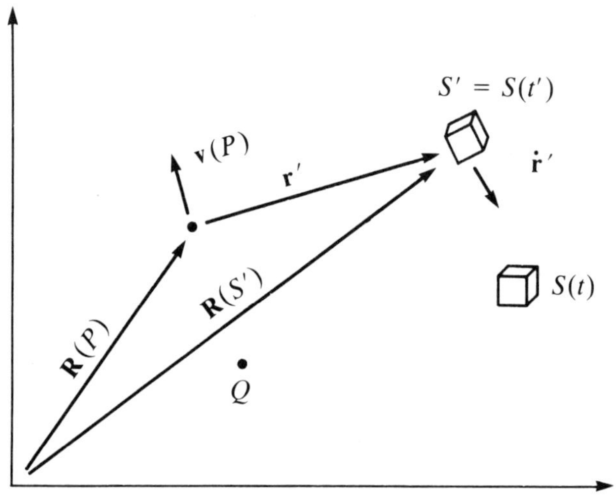

FIG. 6.

seen in Fig. 6, the point P itself has the velocity $\mathbf{v}(P)$ relative to the origin of the $\mathbf{R}$-coordinate system. Therefore, the embedding of the potential $\mathbf{\Phi}^i(P, t)$ into the common coordinate system depends upon $\mathbf{v}(P)$. In the transformation from $\mathbf{r}$-coordinates to $\mathbf{R}$-coordinates, it is necessary to specify $\mathbf{v}(P)$ in addition to $\mathbf{R}(P)$. The velocity of the source point is given by $\dot{\mathbf{r}}' = \mathbf{v}(S') - \mathbf{v}(P)$. One must specify the location $(\mathbf{R}, \mathbf{v})$ in phase space of the unit test particle to be inserted at an arbitrary point in space, $\mathbf{R}(P)$. Upon being embedded in the common coordinate system, the vector retarded potential is denoted $\mathbf{\Phi}^i(\mathbf{R}, \mathbf{v}, t)$.

Conceptually, retarded potentials are simple, being an accounting of actions propagated at a finite velocity and weighted by the inverse of the distance between emitter and receiver. Computationally, retarded potentials are not so simple. A physics based upon retarded potentials as the measure of interaction between bodies is ipso facto relativistic. The integrals for the retarded potentials are first formulated in terms of relative distance, relative velocity, and a retarded time t' particular to an interaction between the receiver and specific emitter. This retarded time is transformed to a common time via $t' = t - r'/c$. The variation of t' over all the past history of the system is thus transformed into a variation of r' over all space. Space and time particular to interactions become intertwined upon imbedding the retarded potential in a common time system.

The physics presented in the preceding chapters tacitly assumed the common $\mathbf{R}$-coordinate system. Fluxes, for example, are given in terms of species velocities $\mathbf{v}_k$ relative to the origin of this common coordinate system. The substantial time derivative $D_{t,k}$, is prescribed relative to this system. And this raises a central problem of physics. How should one choose the origin for the common coordinate system in which one frames dynamics and energetics? This is the problem of the inertial coordinate system. One solves the problem by being partial to the potentials associated with mass. The **R**-coordinate system is chosen so that the physics of the dynamics of mass is simple. A very readable account of the problem of inertial coordinate systems, rich in history and philosophical comment, is given in a short monograph by Sciama (1969). Underlying the problem of finding the inertial coordinates is the fundamental question of the cause of Newton's inertial force $\mathbf{F} = m\mathbf{a}$. The conjecture that it arises from interactions with the "fixed stars" has become known as Mach's principle. Sciama presents a plausibility argument for Mach's principle using an equation giving the force on an electrical charge in terms of first- and second-order time derivations as an analogue for the inertial force on masses.

E. The Scalar Potential for Mass

The scalar potential for mass is commonly called the gravitational potential, or Newton's gravitational potential. The property associated with gravity is mass, and $a^i = \rho$, the density of mass. The normalization factor for the retarded integral is the universal gravitation constant: $K^i = -G$. The minus sign indicates that masses attract one another. The scalar potential for gravity is given by (156) as

$$\phi_g(\mathbf{R}, t) = -G \int_{\Omega} \frac{\rho(\mathbf{r}, t - r'/c)}{r'} dV'. \tag{160}$$

In accord with Mach's principle, the integration is over the whole volume of the cosmos Ω. The scalar potential is embedded in the common coordinate system shown in Fig. 7. The placement of the origin is a matter of computational convenience—say, at the center of the solar system. The origin is at rest with respect to the "fixed stars." Just what this means becomes apparent in the next section, where the vector potential is derived. The motion of the origin is of no consequence for the scalar potential, since it is evaluated statically, with all bodies momentarily at rest.

Space is divided into two regions: (1) local space Ω_L enclosing a region close to the origin in which the potentials of interest are to be evaluated, and (2) cosmic space Ω_C extending outward from the preceding region to the furthest galaxies or an event horizon. Hence, $\Omega = \Omega_L + \Omega_C$. The division of space into two regions is to a great extent arbitrary. One might choose the distance to the nearest star for the radius R_L of the local region and the

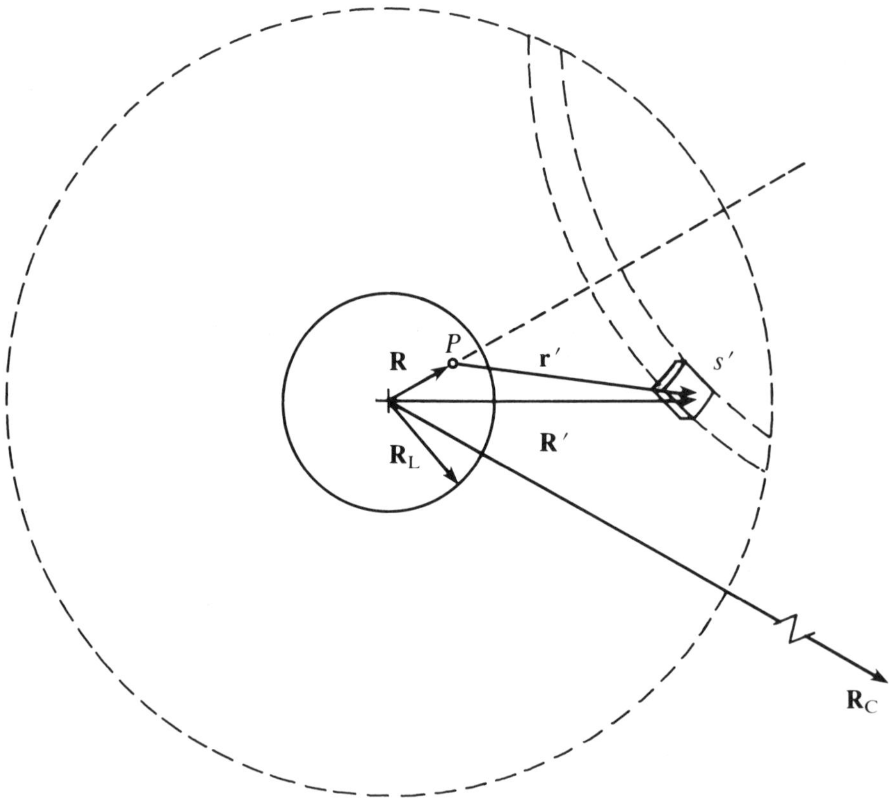

FIG. 7.

velocity of light divided by Hubble's constant as the radius $R_C = c/H$ of the outer edge of the cosmic region. The latter corresponds to what is generally used as the event horizon. There are two possible ways to think of an event horizon: (1) the propagation time for an action from an event at a distance greater than R_C is greater than the age of the cosmos, or (2) the galaxies at distances greater than the R_C are receding with a velocity greater than the propagation velocity c. The local region is very small as compared to the cosmic region since $R_L \cong 4$ light years and $R_C \cong 10^{10}$ light years.

The scalar potential (160) is the usual Newtonian gravitational potential except that it has retarded arguments and includes contributions from Ω_C. It is reasonable to assume that, on the large scale, mass in the cosmos is uniformly distributed. Thus, in the cosmic region Ω_C, mass is distributed with a constant average density ρ. Consider the contribution to the potential from a spherical shell in Ω_C of thickness dR' at a radial distance R'. This contribution to the potential at $\mathbf{R}$ is

$$d\phi_g(\mathbf{R}, t) = -Gdm/R', \tag{161}$$

where $dm = 4\pi\rho R'^2\, dR'$ is the total mass in the shell. This is a well-known result of elementary physics—the gravitational potential at any point inside a spherical shell is a constant, independent of the point's position $\mathbf{R}$. The details of this exercise in geometry are given in Volume 1 of the "Feynman Lectures."

An integration from R_L to R_C gives the total contribution from the cosmic region:

$$\phi_g(\text{cosmic}) = -4\pi G \int_{R_L}^{R_C} \rho R'\, dR' = \text{const.} \tag{162}$$

One sees here that the precise values of R_L, R_C, and ρ are immaterial because the precise value of ϕ_g (cosmic) itself is immaterial. A constant potential gives rise to no forces and can be ignored. For real-potentials, the gradients of the ϕ^i appear in the dynamics and energetics, but not the ϕ^i. The density, ρ was purposely left inside the integral in (162) to show that the conclusion, $\phi_g(\text{cosmic}) = \text{const}$, holds under the less restrictive assumption $\rho = \rho(R')$.

In light of (162) and the fact that a real-potential does not have gauge, the scalar potential for gravity can be calculated by an integration over only the local region Ω_L. Furthermore, the distances in a laboratory or even in the solar system are small enough for retardation to be ignored (up to relativistic effects, which are of the order of v^2/c^2). Thus (160) can in that case be reduced to the Newtonian potential,

$$\phi_g(\mathbf{R}, t) = -G \int_{\Omega} \frac{\rho(\mathbf{r}, t)}{r}\, dV. \tag{163}$$

F. The Vector Potential for Mass

The common coordinate system will be chosen so that the vector potential for mass is a simple expression. This might be phrased in a different manner. In light of the development of the unified dynamics given in the preceding five sections, it is imperative to demonstrate that the assumed vector potential for mass, $\mathbf{\Phi}_g = \mathbf{v}$, can indeed be derived from a retarded integral. This will entail a special choice for the common coordinate system. The choice of an origin at rest (or in uniform motion) with respect to the "fixed stars" comes as no surprise; for centuries this has set the coordinates for analytical mechanics.

The task at hand is to calculate the vector potential (159) at a point $P(\mathbf{R}, \mathbf{v}, t)$ in phase space. In keeping with Mach's principle, the integration is over the whole cosmos, $\Omega = \Omega_L + \Omega_C$. A unit test particle at the point $P(\mathbf{R}, \mathbf{v}, t)$ moves with the velocity $\mathbf{v}$ with respect to the origin of the $\mathbf{R}$-coordinates. It thus moves with respect to the "fixed stars" with the velocity $\mathbf{v}$. As noted in Section VII.D, the velocity of the source point S' relative to the point $\mathbf{P}$ is given by $\dot{\mathbf{r}}' = \mathbf{v}(S') - \mathbf{v}(P)$. Since the "fixed stars" are fixed by virtue of $\mathbf{v}(S') = 0$, the relative velocity that appears in the integral (159) for the vector potential is $\dot{r}' = -\mathbf{v}(P) = -\mathbf{v}$. An observer stationed on a unit test particle at $P(\mathbf{R}, \mathbf{v}, t)$ sees the "fixed stars" move past him at a velocity $-\mathbf{v}$.

The category "fixed stars" includes the galaxies. The radial distances from the test particle to the galaxies are large, to say the least, but because of the overwhelming amount of mass in Ω_C, the contribution to the vector potential from the local region Ω_L can be ignored (up to relativistic effects). Under the assumption that mass in Ω_C is uniformly distributed with a constant density ρ, retardation is no longer a factor. Furthermore, since the distance to the point P is much smaller than the distance r' to the galaxies, the approximation $\mathbf{R} = \mathbf{r}'$ (see Fig. 7) allows the integration of (159) to be performed directly in $\mathbf{R}$-coordinates. The vector potential (159) for mass is thus given by

$$\mathbf{\Phi}_g(\mathbf{R}, \mathbf{v}, t) = \frac{G\rho\mathbf{v}(t)}{c^2} \int_{\Omega_C} \frac{1}{R}\, dV. \tag{164}$$

Before proceeding, a few remarks should be made. The minus sign from $K^i = -G$ has been cancelled by the minus sign of the relative velocity of the galaxies with respect to $P(\mathbf{R}, \mathbf{v}, t)$. If the origin of the $\mathbf{R}$-coordinates moved uniformly with respect to the "fixed stars" with some constant velocity, the vector potential (164) would be augmented by a constant vector term. This would not affect the dynamics because only $D_{t,k}\mathbf{\Phi}_g$ appears there, not $\mathbf{\Phi}_g$ alone. The substantial time derivation of a constant vector is zero; therefore a term in $\mathbf{\Phi}_g$ arising from a uniform motion of the origin can be ignored. Finally, the assumption that the galaxies are at rest with respect to each other can be relaxed. In an expanding universe the expansion is symmetrically outward from any point in space. Hence from the vantage point of the given test particle, for each galaxy at the relative position $\mathbf{r}'$, receding with a velocity $\dot{\mathbf{r}}'(r')$ relative to the particle, there is by symmetry and uniformity another at $-\mathbf{r}'$ receding with velocity $-\dot{r}'(r')$. Therefore the mass-flux contributions to the vector potential (159) arising from the expansion cancel, giving once again the potential (164).

The goal of this section has been to demonstrate that the identification $\mathbf{\Phi}_g = \mathbf{v}$ could be derived, in accord with Mach's principle, from a retarded integral of the fluxes of galactic masses in motion relative to a test particle. This was accomplished in (164), up to a constant factor. That is, $\mathbf{\Phi}_g = I\mathbf{v}$,

Table II

Physical Constants

$G = 6.67 \times 10^{-11}\ \mathrm{m^3\ kg^{-1}\ sec^{-2}}$
$H = 80\ \mathrm{km\ sec^{-1}\ Mpc^{-1}}$
$1\ \mathrm{Mpc} = 3.09 \times 10^{19}\ \mathrm{km}$
$\rho = 3 \times 10^{-28}\ \mathrm{kg\ m^{-3}}$

where I is a constant that should be unity. By (164),

$$I = \frac{4\pi G\rho}{c^2} \int_{R_L}^{R_C} R\, dR. \tag{165}$$

As before, the edge of the cosmic region will be determined by the event horizon $R_C = c/H$. Since $R_L \ll R_C$, the integration can be simplified by setting $R_L = 0$, giving

$$I = 2\pi G\rho/H^2. \tag{166}$$

The values of the physical constants appearing in this equation are given in Table II. The value of the constant factor in the calculated vector potential for mass is, therefore,

$$I = 0.02, \tag{167}$$

which is rather different from the desired $I = 1.00$. Using the commonly cited value of Hubble's constant of $H = 55\ \mathrm{km\ sec^{-1}\ Mpc^{-1}}$ would help only slightly, giving $I = 0.04$. The trend, however, is toward increasing values of H, with $H = 80\ \mathrm{km\ sec^{-1}\ Mpc^{-1}}$ as a best average value (Heidmann, 1977). Another way to improve the situation would be to assume that the cited value of ρ is too small. This value is based upon the distribution of the bright matter in the universe, the luminous galaxies seen on photographic plates. Given a count of galaxies in a representative region of space and a reasonable figure for the mass of an average galaxy, the mean density ρ can be calculated. The value of ρ in Table II is determined in this manner. If Mach's principle is the explanation for inertia and if inertia can be calculated from a retarded potential, then, since all the experience of mechanics dictates that $I = 1.00$, some mass must be missing. Indeed, a fair amount of mass is missing.

The problem of "missing mass" has a long history in cosmology and is still a central issue. It occurs in a context slightly different than, but related to, the preceding problem of inertia. One question immediately raised by Einstein's theory of general relativity was whether the universe is closed or open. Mass curves space–time; and with a sufficient density of mass, the universe would be curved into a closed region. With increasing evidence for a big-bang genesis of the universe, the question of closure has once again become current. The Einstein closure criterion is that

$$\rho \geq \rho_{\mathrm{E}} = (3/8\pi G)H^2. \tag{168}$$

For $H = 80$ km sec^{-1} Mpc^{-1}, the critical density for closure is $\rho_{\mathrm{E}} = 1.2 \times 1.0^{-26}$ kg m^{-3}. This is much greater than the density of bright matter cited in Table II. Either the universe is open or a lot of mass is "missing."

The similarity between (166) and (168) allows an intriguing calculation. Assume that the derivation that resulted in (166) is valid. From the necessity of having $I = 1.00$, one can calculate the density ρ_I required to provide the observed magnitude of inertia. However, it is perhaps more interesting to use the fact that $I = 1.00$ to determine whether or not the universe is closed. Eliminating the physical constants G and H from (166) and (168) gives

$$\rho_I = (1 + \tfrac{1}{3})\rho_{\mathrm{E}}. \tag{169}$$

Therefore, under the premise that (166) is valid, it must be concluded that the universe is closed. However, the precision of (169) must be taken with more than a little bit of caution. Or, rather, the whole of the preceding discussion must be taken in the proper spirit.

This digression into cosmological matters is presented to show that it is plausible to derive inertia from a retarded potential. It was desired to show that the vector potential for gravity could be obtained from an integral over mass flux just as the vector potential for electricity can be obtained from an integral of charge flux. It should be noted that none of the thermodynamics developed in the preceding chapters depends upon how the various potentials are to be determined. To make the connection to classical physics, it was demonstrated in Section III that the dynamics of electromagnetism and of hydrodynamics follow immediately from the four-vector gradient operating upon $\mathbf{\Phi}_{\mathrm{e}} = \mathbf{A}$ and $\mathbf{\Phi}_{\mathrm{g}} = \mathbf{v}$. The argument presented above to show that the identification $\mathbf{\Phi}_{\mathrm{g}} = \mathbf{v}$ is plausible does lead to an intriguing result: the cosmic mass density required in this simple model for inertia is, to all extents and purposes, equal to that required to close the universe.

The goal of this work is to develop thermodynamics within the context of classical field theory, not to explore the more arcane science of cosmology. Nevertheless, it seems unfair to cut short this digression without addressing the obvious question: Is there any experimental evidence for the existence of the missing (invisible) matter? Measurements of velocities in different regions of rotating galaxies indicate large drag forces arising, presumably, from significant amounts of dark matter. This is substantiated by radio observations at 21 cm that show great halos around the galaxies. The conclusion that bright matter may constitute as little as 10% of the mass of a galaxy has, among other things, prompted a reexamination of models for the formation of galaxies, for example, that of White and Rees (1978). Another method to determine the average density of galaxies is to measure the gravitational influence of the Virgo supercluster of galaxies upon the motion of the Local group due to the expanding universe. The observed perturbation indicates an average density of 40–70% of that required for closure, (Davis *et al.*, 1980; Aaronson *et al.*, 1980).

G. Electrodynamics

The electrostatic potential is given by the retarded integral (156) with a normalization constant $K^i = \frac{1}{4}\pi\varepsilon_0$, where ε_0 is the permittivity of free space. In this general discussion there is no need to consider dielectric media. That the normalization constant is positive means that charges of like sign repel one another, whereas charges of opposite sign attract one another. The property density that appears in the integrand is charge density; that is,

$$a^i(\mathbf{r}', t - r'/c) = \bar{Z}_k C_k(\mathbf{r}', t - r'/c). \tag{170}$$

The vector potential $\mathbf{A}$ is given by the retarded integral (159) with the same normalization factor as for the scalar potential. The flux that appears in the integrand is the electrical current flux; that is,

$$\mathbf{J}^i(\mathbf{r}', t - r'/c) = \bar{Z}_k C_k(\mathbf{r}', t - r'/c)\dot{\mathbf{r}}'_k. \tag{171}$$

The domain of integration for both retarded integrals is the local region Ω_L. Since the world on the large scale is electrically neutral, the integrals over the cosmic region are zero.

H. An Electrical Model for Gravitational Fine Structure

In Section VII.F the retarded vector potential for mass was calculated by an integral over the whole cosmos, in compliance with Mach's principle. Assuming that the contribution to the integral from mass fluxes in the local region was negligible, it was found that $\mathbf{\Phi}_g \cong \mathbf{v}$, in accord with Newton's law of inertia. If, however, the interactions arising from the motions of masses in the local region are not neglected, the complete vector potential is

$$\mathbf{\Phi}_g = \mathbf{\Phi}_g(\Omega_L) + \mathbf{\Phi}_g(\Omega_C) \tag{172}$$

since Ω_L and Ω_C are disjoint regions.

In Section VII.E it was demonstrated that the contribution to the retarded scalar potential from the cosmic region was zero, by reason of a uniform and symmetrical distribution of mass on the cosmic scale. Since $\mathbf{\Phi}_g(\Omega_C) = \mathbf{v}$, the exact expressions for the vector and scalar potentials for mass are

$$\boldsymbol{\Phi}_g = \boxed{-\frac{G}{c^2}\int_{\Omega_L} \frac{\mathbf{J}_g(\mathrm{r}', t-r'/c)}{r'}\,dV'} + \boxed{\mathbf{v}} \tag{173}$$

$$\phi_g = \boxed{-G\int_{\Omega_L} \frac{a_g(\mathbf{r}', t-r'/c)}{r'}\,dV'} \tag{174}$$

where $\mathbf{J}_g$ is mass flux and a_g is mass density.

The dynamics of gravitating bodies is contained in the preceding two potentials. From the viewpoint of the underlying mechanisms, (173) represents interactions arising from relative motion, whereas (174) represents interactions that can be characterized by relative position. The first is a vector and the second is scalar—different in cause and different in tensorial order. The associated dynamics is governed by the space- and time-related fields $-\nabla\phi_g$ and $-D_t\boldsymbol{\Phi}_g$, where $\boldsymbol{\Phi}_g$ now includes the fine structure given by the integral over Ω_L in (173).

From the practical viewpoint of calculation, there is much to be gained by ignoring the underlying mechanisms and considering a dynamics based upon the grouping of effects implied by the boxes drawn around the terms of (173) and (174). The second box contains only one term, and this term gives rise to the Newtonian inertial force. The first box contains two retarded integrals (vector and scalar), both evaluated in the local region. These two integrals are in form exactly like those describing the dynamics of electrically charged bodies. The conclusion, therefore, is that a complete dynamics for gravitating bodies can be provided by an electrodynamic model augmented by an inertial force.

Electrical models for gravity have a long history. Tisserand, for example, in 1872 used Weber's force law to calculate the precession of the perihelion of Mercury, a phenomenon that could not be accurately predicted by Newton's laws. Lévy (1890), improving upon Tisserand's attempt, successfully predicted the observed precession. It is interesting to note that Tisserand and Lévy saw the additional force terms from the electrical model as corrections to the static potential, whereas from the viewpoint of the preceding discussion they are, in fact, corrections to the inertial force.

I. The Wave Equations and the Lorentz Gauge

The following brief remarks are included for the sake of completeness. As far as the formal development of the dynamics and energetics of complex real systems, they lie beyond the scope of this book. Upon differentiation with respect to space and time, with proper regard for the retarded arguments, the integrals (156) and (159) for the scalar and vector potentials can be transformed into the following equivalent equations:

$$\nabla^2 \mathbf{\Phi}^i - \frac{1}{c^2}\frac{\partial^2 \mathbf{\Phi}^i}{\partial t^2} = -\frac{4\pi K^i}{c^2}\mathbf{J}^i \tag{175}$$

and

$$\nabla^2 \phi^i - \frac{1}{c^2}\frac{\partial^2 \phi^i}{\partial t^2} = -4\pi K^i a^i. \tag{176}$$

This result can be summarized by the following theorem.

THEOREM. The scalar and the vector parts of real-potentials are given as solutions to the wave equations (175) and (176).

By Theorem (134) the law of conservation of properties conjugate to real-potentials is

$$\partial a^i/\partial t = -\nabla \cdot \mathbf{J}^i. \tag{177}$$

This equation relates the time behavior of a^i with the spatial behavior of $\mathbf{J}^i$. With a^i given by (175) and $\mathbf{J}^i$ given by (176) this equation results in

$$\partial \phi^i/\partial t = -c^2 \nabla \cdot \mathbf{\Phi}^i. \tag{178}$$

In electrodynamics this equation is known as the Lorentz gauge for **A**. In the present context (178) is not a gauge by assumption or choice, but rather is a mathematical consequence of basing the physics of interaction upon retarded integrals—in effect, upon actions propagated with a finite velocity.

VIII. PSEUDO-POTENTIALS

A. Information Content

The division of the system into species of particles at a given level in the hierarchy of physical descriptions presupposes the existence of measuring devices capable of resolving the continuum into those designated categories. Or, equivalently, the engines proposed to extract work from the system must posses sufficient discrimination to distinguish the separate species of particles.

A measuring device is, in essence, an engine that energetically couples to the system in such a manner as to produce a minimal perturbation of the system. The point is that identity implies measurement, and measurement, in turn, implies energy available to do work upon a meter.

Consider a solution containing several species of molecules. There is no need for a distinction between solvent and solute molecules. The solution is a message and the members of the various species constitute the symbols. Suppose a probe from a measuring device is put into the solution, and a molecule is recognized as belonging to a certain species. Some information about the solution has been gained; a symbol has been read. A quantitative measure of the information content of a molecule of species k as a symbol in a message is given by information theory as

$$I_k = \alpha \ln p_k \tag{179}$$

where p_k is the *a priori* probability of the event and α is a constant to be determined later. The probability of finding at random a molecule of species k in the solution is just the mole fraction, X_k; that is,

$$p_k = X_k = C_k/C, \tag{180}$$

where $C = \Sigma\, C_k$ is the total concentration of molecules at that particular location in the solution. Since the concentration is a function of space and time, the information content (179) is also.

The average information per symbol I is simply the mean value of the information carried by the various symbols; hence

$$I = p_k \alpha \ln p_k = X_k \alpha \ln X_k. \tag{181}$$

At any point in the solution (i.e., the continuum) the concentration of symbols is C; therefore the volume density of information is

$$\rho(I) = CI = C_k \alpha \ln X_k. \tag{182}$$

B. The Informational Potential

Information is energy, and a proper choice of α puts the information density (182) into energy units. In general, the density of energy in the continuum is, by (16), simply

$$u = C_k u_k, \tag{183}$$

where the scalar potential energy per mole of species k is, by (1), given as the mapping

$$u_k = A_k^i \phi^i. \tag{184}$$

Written in terms of this formalism, the energy associated with information must be represented as

$$\rho(I) = u(I) = C_k A_k^i(I)\phi^i(I), \tag{185}$$

where the argument I indicates that the parameters and variables are those related to information. A comparison of (182) with the formal statement (185) shows that

$$A_k^i(I)\phi^i(I) = \alpha \ln X_k. \tag{186}$$

The resolution of (186) into the desired informational potentials μ_k comes from a consideration of the indices. The index i must stand in a one-to-one correspondence to the index k if (186) is to make sense mathematically. That is, there is a potential $\phi^i(I)$ for each molecule species k. Furthermore, the scale factors are identically equal to one (see Section VIII.D). The resolution, then, is as follows:

$$A_k^i(I) = 1 \tag{187}$$

$$\phi^i(I) = \alpha \ln X_k \equiv \mu_k. \tag{188}$$

Note that in any sum over the index i—say in $u = A_k^i\phi^i$—one must remember to include the informational potentials, $\mu_k = \phi^i(I)$. As pointed out before, pseudo-potentials arise from necessary adjustments to insure congruence between the p-space and the s-space descriptions. A mixing (or sharing) of indices in that case in not unexpected.

C. Energy Equivalence of Information

The classical device used to decipher the message in a solution is an osmometer made by closing off the end of a thistle tube with a semipermeable membrane. A small volume of sugar solution is put in the tube. The end of the tube closed off by the membrane is then slightly submerged in a beaker of pure water (or of a more dilute solution). The membrane is said to be ideally semipermeable if it can make a perfect distinction between solute and solvent molecules—water can permeate but sugar cannot. Water flows from the beaker into the thistle tube. The potential energy of information is converted to gravitational potential energy (expressed as a pressure difference across the membrane). At equilibrium the driving force upon the water molecules has become zero. The potential energy density for water, u_w is the same on both sides of the membrane; that is,

$$(C_w \bar{V}_w p)^b = (C_w \bar{V}_w p + C_w \alpha \ln X_w)^t, \tag{189}$$

where b and t refer to points immediately at the solution–membrane interface in the beaker and in the tube. In (189) it was implicitly assumed that the membrane was thin enough that the drop in gravitational potential energy across it could be neglected. For convenience, it is assumed that pure water is in the beaker. In that case $X_{\mathrm{w}}^{\mathrm{b}} = 1$ and $\mu_{\mathrm{w}}^{\mathrm{t}} = 0$ (thus, like the other pseudo-potentials the informational potential has gauge). If the solution in the thistle tube is dilute, then on the tube side of the membrane

$$X_{\mathrm{w}} = \frac{C_{\mathrm{w}}}{C_{\mathrm{w}} + C_{\mathrm{s}}} \cong \frac{1}{1 + \bar{V}_{\mathrm{w}} C_{\mathrm{s}}} \cong 1 - \bar{V}_{\mathrm{w}} C_{\mathrm{s}}, \tag{190}$$

since $\bar{V}_{\mathrm{w}} C_{\mathrm{w}} \cong 1$ for dilute solutions. Furthermore, since $\bar{V}_{\mathrm{w}} C_{\mathrm{s}} \ll 1$, the logarithm in (189) can be expanded to give the well-known formula

$$\Delta p = \alpha C_{\mathrm{s}}. \tag{191}$$

The pressure difference Δp across the membrane can be calculated from the height to which the solution has risen in the tube. It is found that α depends upon temperature and is given by $\alpha = RT$, where R is the universal gas constant. Therefore the informational potential for particles of species k is

$$\mu_k = RT \ln X_k. \tag{192}$$

Had one considered a semipermeable membrane separating two mixtures of ideal gases, Dalton's law of partial pressures ($p_k = RTC_k$ where here p_k is a partial pressure) would have provided a theoretical derivation of the value of α in (191).

D. Concentration as a Property

The informational potential (192) is given as a direct measure of the energy per mole of species k conferred upon the molecules by virtue of being identifiable. This stands in contrast to the units of all other potentials, which are mapped to potential energies by scale factors having units such as mass per mole or volume per mole. The scale factors have units of property i per mole of species k. The property conjugate to the informational potential is, as will be seen, the species concentration C_k.

Consider the potential (192) again. Since the scale factor for information was chosen to be unity, the potential is directly in potential energy units. In accord with the general mapping $a^i = A_k^i C_k$, the property associated with the informational potential is $C_k = A_k^i C_k$, but $A_k^i = 1$, so

$$C_k = 1 C_k \tag{193}$$

since here $A_k^i = 1$. Concentration is thus a property. As mentioned in an earlier chapter, the species concentration is the sole property in the p-space description. By the mapping (193) it is also a property in s-space. Finally, it should be noted that, in general, a scale factor A_k^i maps a property in p-space (i.e., C_k) to a property in s-space (i.e., a^i).

All properties have equations of continuity. Obviously, the equation of continuity for the property conjugate to the informational potential is the familiar

$$\frac{\partial C_k}{\partial t} = -\nabla \cdot \mathbf{J}_k + \sigma_k. \tag{194}$$

This introduces nothing new, but it does demonstrate that the formalism presented in Section II encompasses the concept of a potential based upon identity and information content.

If, in (186) $\alpha = RT$ had been chosen as the scale factor for information, the associated property would have been RTC_k instead of (193). This is immediately ruled out because it creates a confusion between property and potential. Even worse, there obviously could be no associated equation of continuity such as (194).

E. Activity Coefficients

The derivation of the informational potential (192) assumed that the identity of a particle was uncompromised by its inclusion in the continuum. In point of fact, a solution for which (194) holds is called an ideal solution; in the real world, particle identity is much affected by the proximity of other molecules. A solute molecule may to some extent lose its identity as a single identifiable particle by forming loosely bonded polymers. The most common cause of a loss of particle identity is the forming of weakly bonded complexes with other species, either solvent or solutes. These bonds are weak as compared with the chemical bonds involved in aggregation processes. In aggregation the bonds are so strong that the component particles (reactants) completely lose their identity in the creation of a product having its own individual identity.

The possible manifestations of weak bonding are legion, and their theoretical characterization is for the most part highly impractical. This is one of the most compelling reasons for the physical chemistry of solutions to be a semiempirical science. To make the ideal potential (192) conform to the realities of experience, it is adjusted by a factor that may depend upon the physical state and the chemical composition of the continuum. This factor γ_k is called the activity coefficient, and the adjusted potential is given as

$$\mu_k = RT \ln \gamma_k X_k. \tag{195}$$

F. Field Equations for Pseudo-Potentials

Real-potentials are a measure of the intensity of interactions between particles and are evaluated either by retarded integrals or wave equations. The wave equations (175) and (176) are called time-reversible because the time derivative is to the second order. The corresponding equations for the determination of pseudo-potentials are, in contrast, time-irreversible. They contain a first-order time derivative. Before turning to the temperature potential T, consider the informational potential. Since μ_k is determined by the concentrations C_k, (194) is, in effect, the field equation for the informational potential. Obviously it is irreversible.

In certain situations with no viscous effects and where inertial effects can be neglected, the complex dynamics (112) or (113) can be replaced by the linear approximation

$$\mathbf{J}_k = -L_k \nabla \mu_k. \tag{196}$$

For ideal solutes in dilute solutions, this flux equation can be written in a familiar form as

$$\mathbf{J}_k = -D_k \nabla C_k, \tag{197}$$

where the diffusion coefficient is defined by

$$D_k = RTL_k/C_k. \tag{198}$$

For the sake of illustration, consider the continuity equation (194) for solutes obeying (197). One obtains the lovely field equation

$$\partial C_k/\partial t = -\nabla \cdot (-D_k \nabla C_k) + \sigma_k = D_k \nabla^2 C_k + \sigma_k. \tag{199}$$

For simplicity it was assumed that the diffusion coefficients are constants (often the case for dilute solutions).

The continuity equation for concentration, (194), stands as the field equation for the informational potential whether or not a linear approximation such as (196) holds. Equation (196) in the form (197) is called Fick's diffusion equation. The study of linear approximations of the type (196) has become known as irreversible thermodynamics (see, e.g., deGroot and Mazur (1963) or Katchalsky and Curran (1965)). There it is assumed that any flux is a linear function of all the driving forces. This is referred to as coupling and is expressed by

$$\mathbf{J}_i = L_{ik} \mathbf{F}_k(u). \tag{200}$$

Following the arguments of irreversible thermodynamics, by using the concept of force developed in this essay, the preceding flux equation can easily be generalized to

$$\mathbf{J}_i = L_{ik}\mathbf{F}_k, \tag{201}$$

where $\mathbf{F}_k$ is the net force per mole of species k. This formally incorporates inertial effects but, as far as solving dynamical problems, is calculationally impractical. This subject will not be pursued further as it partakes more of phenomenology than of physics (see Richardson, Louie, and Swaminathan (1982) and Richardson and Louie (1983)). This is not to say that the linearizations of irreversible thermodynamics may not often provide valuable insights into the behavior of a large class of dynamical systems.

G. The Caloric Equation

The local density of scalar potential energy is represented in s-space by a bilinear form (84) in properties and potentials: $u = a^i\phi^i$. The thermal contribution to this potential energy is the product of entropy density and temperature: u (thermal) $= sT$. This thermal contribution to scalar potential energy can increase by an increase in entropy and/or an increase in temperature. The greater part of Section V was devoted to the derivation of an equation of continuity for entropy—that is, the determination of $\partial s/\partial t$. It remains to find $\partial T/\partial t$.

The local temperature of the continuum changes because of (1) a divergence in the heat flux $\mathbf{J}_q$, (2) heat produced by dissipative processes, and (3) heat released (or absorbed) in chemical reactions. The rate of change in temperature is thus given by the caloric equation:

$$C_V\,\partial T/\partial t = -\nabla\cdot\mathbf{J}_q + \delta + u_R\,\partial\xi_R/\partial t \tag{202}$$

where C_V is the specific heat at constant volume of the continuum. By (98) and (99) this can also be written as

$$C_V\,\partial T/\partial t = -\nabla\cdot\mathbf{J}_q + T\sigma. \tag{203}$$

The caloric equation is, in effect, the field equation for the thermal potential T.

Analogous to Fick's equation (197), there is a linear approximation for heat flux that should be mentioned. In certain restricted situations the heat flux is given simply as

$$\mathbf{J}_q = -\kappa\nabla T \tag{204}$$

where κ is the thermal conductivity of the continuum. This is Fourier's equation. In this case, the field equation (203) for the thermal potential T becomes

$$\partial T/\partial t = (\kappa/C_V)\nabla^2 T + T\sigma/C_V. \tag{205}$$

Hence the field equations for the two pseudo-potentials μ_k and T have more or less the same form, (199) and (105), for systems wherein the associated fluxes, (196) and (204) are given by linear phenomenological equations. It must, however, be emphasized that there is no general canonical form for the field equations for pseudo-potentials. The existence and character of the pseudo-potentials depend in general upon the hierarchial structure of the given system and upon how one chooses the level in the hierarchy at which the p-space description is framed.

H. Pressure

It is appropriate to end this chapter (and hence this essay) with a discussion of pressure. But, in truth, there will be no discussion of pressure. The concept of pressure one has from elementary thermostatics and hydrodynamics suffices for understanding the formal development of energetics and dynamics presented in this essay. The next step—the investigation of pressure—is not a simple one to be contained in a section or even a chapter. It is a step into another discipline: the mechanics of continua and deformable bodies.

A discussion of pressure leads immediately to the more general problem of the stress tensor. From there it turns to Hooke's law, which proposes that the components of the stress tensor are linear functions of the components of the strain tensor. Thence a whole science develops, and with it an energetics that must account for such things as thermoelastic materials. For a start in that direction, the reader is referred to Volume 2 of the "Feynman Lectures."

ACKNOWLEDGMENTS

I am deeply indebted to my Polish colleague Stanislaw Miekisz for the genesis of this monograph lies in two essays (in the proper sense of the word) on field thermodynamics that we published in the *Bulletin of Mathematical Biology* (1976, *38*: 479–496, and 1978, *40*: 301–318). I thank Bob Rosen, Aloisius Louie, and all of my other friends in the Red House for helpful suggestions and enthusiastic support during the slow evolution of this work.

REFERENCES

Aaronson, M., Mould, J., Huchra, J., Sullivan, W. T., III, Schommer, R. A., and Bothun, G. D. (1980). A distance scale from the infrared magnitude/HI velocity-width relation, III. The expansion rate outside the local supercluster. *Astrophys. J.* **239**, 12–37.

Davis, M., Tonry, J., Huchra, J., and Latham, D. W. (1980). On the Virgo supercluster and the mean mass density of the universe. *Astrophys. J.* **238**, L113–L116.

deGroot, S. R. and Mazur, P. (1963). Non-equilibrium Thermodynamics. North-Holland, Amsterdam.

Feynman, R. P., Leighton, R. B. and Sands, M. (1963) "The Feynman Lectures on Physics" (in 3 volumes). Addison-Wesley, Reading, Massachusetts.

Heidmann, J. (1977). The Expansion of the universe in the frame of conventional general relativity. "Cosmology, History, and Theology" (W. Yourgrau and A. D. Breck, eds.). Plenum Press, New York.

Katchalsky, A. and Curran, P. F. (1965). "Nonequilibrium Thermodynamics in Biophysics." Harvard University Press, Cambridge."

Lévy, M. M. (1890) Sur l'application des lois électrodynamiques au mouvement des planétes. *C. R. Séances Acad. Sci.* **CX**, 545–551.

Lorentz, H. A. (1904) "Electromagnetic Phenomena in a System Moving with any Velocity Less than that of Light." Reprinted in "The Principle of Relativity" by Lorentz, H. A., Einstein, A., Minkowski, H., and Weyl, H. with notes by A. Sommerfeld (1952). Dover, New York.

Richardson, I. W., Louie, A. H., and Swaminathan, S. (1982). A phenomenological calculus for complex systems. *J. Theor. Biol.* **94**, 61–76.

Richardson, I. W., and Louie, A. H. (1983). Projections as representations of phenomena. *J. Theor. Biol.* **102**, 199–223.

Rosen, R. (1978) "Fundamentals of Measurement and Representation of Natural Systems." North-Holland, New York.

Sciama, D. W. (1969) "The Physical Foundations of General Relativity." Doubleday, Garden City, New Jersey.

Tisserand M. F. (1872). Sur le mouvement des planètes autour du soleil, d'après la loi électrodynamique de Weber". *C. R. Séances Acad. Sciences* **XXV**, 760–763.

White, S. D. M., and Rees, M. J. (1978). Core condensation in heavy halos: A two-stage theory for galaxy formation and clustering. *Mon. Not. R. Astron. Soc.* **183**: 341–358.

Whittaker, E. (1973) "A History of the Theories of Aether and Electricity" (2 volumes). Humanities Press, New York.

Wigner, E. P. (1967) Invariance in physical theory. *In* "Symmetries and Reflections." Indiana Univ. Press, Bloomington.

2

Categorical System Theory

A. H. Louie[†]

Department of Physiology and Biophysics
Dalhousie University
Halifax, Nova Scotia

[†] Present address: 86 Dagmar Ave., Vanier, Ontario. No longer affiliated with Dalhousie University.

ISBN 0-12-597280-6

LIST OF SPECIAL SYMBOLS

The symbols listed below are followed by a brief statement of their meaning. The standard set-theoretic symbols are not included here.

$\mathbf{C}$	General category
$\mathbf{C}(A, B)$	Hom set
$f: A \to B$	Morphism
$g \circ f$	Composition of morphisms
dom f	Domain
1_A	Identity morphism
f^{-1}	Inverse morphism
$A \cong B$	Isomorphic objects
Ens	Category of sets
B^A	$= \mathbf{Ens}(A, B)$
Gp	Category of groups
Top	Category of topological spaces
Mon	Category of monoids
Ab	Category of abelian groups
$F: \mathbf{A} \to \mathbf{B}$	Functor
$\mathbf{C}^{\mathrm{op}}$	Dual category
f^{op}	Opposite morphism
h^A	Covariant hom-functor
h_A	Contravariant hom-functor
Cat	Category of (small) categories
$I_{\mathbf{A}}$	Identity functor
$F^{-1}: \mathbf{B} \to \mathbf{A}$	Inverse functor
$\alpha: F \to G$	Natural transformation
$\mathbf{B}^{\mathbf{A}}$	Functor category
ΠA_i	Product
$\pi_i: A \to A_i$	Projection
Σ	Statement
Σ^{op}	Dual statement
$\amalg A_i$	Coproduct
$\iota_j: A_j \to A$	Injection
S	Set of states
$f, g, h, \ldots$	Observables
$\mathbb{R}$	Real numbers
R_f	Equivalence relation induced by f
$f \sim g$	iff $R_f = R_g$, algebraic equivalence
$f \leq g$	iff $R_f \subset R_g$, refinement
$\mathbf{S}$	Category of formal systems
(S, F)	Formal system
F	Set of observables
0	Constant observable
$\phi, \psi, \chi, \ldots$	**S**-morphisms
R_G	$= \cap R_f : f \in G$
eq(f, g)	Equalizer
coeq(f, g)	Coequalizer
$f: A \rightarrowtail B$	Monomorphism
χ_B	Characteristic function
$f: A \twoheadrightarrow B$	Epimorphism
$(S_1, F_1) \leq (S_2, F_2)$	Partial order in **S**
M	Meter
$\mathscr{A}(S)$	Group of automorphisms on S
D	Category of dynamical systems
T	Dynamics
t	Time
$a_x[T], b_x[T]$	Bounds of T at x
$\bar{\mathbb{R}}$	$= [-\infty, +\infty]$, extended real numbers
I_S	Trivial dynamics
y_x	Solutions of a dynamics
$y_x(a_x, b_x)$	Trajectories of a dynamics
T_t	Translations of a dynamics
(S, D)	Dynamical system
D	Set of dynamics
$\phi, \psi, \chi, \ldots$	**D**-morphisms
$\phi \times 1_{\mathbb{R}}$	$:(x, t) \mapsto (\phi x, t)$
$\mathbb{N}$	Natural numbers
$(\mathscr{E}, \mathscr{M})$	Image factorization system
$\mathscr{R}(S)$	Lattice of equivalence relations on S
R_T	Conjugate equivalence relation
L_T	Conjugate lattice

$\mathbb{Z}$	Integers
$\langle T \rangle$	$= \{T^n : n \in \mathbb{Z}\}$, discrete dynamics
τ	Topology
d_f	f-induced pseudometric
$E: N \to S$	Modelling relation
$\mathbf{N}$	Category of natural systems
(S, F, D)	Natural system
Φ, Ψ	Linkage relations
$(S_1, F_1, D_1) \leq (S_2, F_2, D_2)$	Partial order in $\mathbf{N}$
$\mathbf{L}$	Organism
L_i	$\mathbf{L}$-objects
ϕ_{ji}	$\mathbf{L}$-morphisms
$\mathbf{K}$	Structured category
$\mathbf{K}(X)$	$\mathbf{K}$-structures on X
(X, σ)	$\mathbf{K}$-structure
$\mathbf{K}(\sigma, \tau)$	$\mathbf{K}$-admissible $\mathbf{C}$-morphisms
$\mathbf{K}$	Category associated with $\mathbf{K}$
S	Organismic set
e_i	Element of organismic set
S_i^a	Activities of e_i
S_i^p	Products of e_i
S^a	Activities of S
S^p	Products of S
S_1, S_2, S_3	Specialized subsets of S
t_i	Survival time of e_i
S_R	Product-induced relations
$\mathbf{O}$	Organismic category
V	Vector (inner product) space
V^*	Dual vector space
$\mathbf{F}, \mathbf{F}_i$	Contravariant vectors
$\mathbf{a}, \mathbf{a}^i$	Covariant vectors
$\mathbf{R} = \mathbf{aF}$	Dyad
$T_1^1(V)$	Tensor space of type (1, 1) over V
$\mathbf{R} = \Sigma_i \mathbf{a}^i \mathbf{F}_i$	Dyadic (= response tensor)
$\mathbf{R} : \mathbf{S}$	Double dot product
$[\mathbf{a}^1, \ldots, \mathbf{a}^m]$	Description space
$\lvert \mathbf{R} \rvert$	$= (\mathbf{R} : \mathbf{R})^{1/2}$, norm of $\mathbf{R}$
L^2	Square-integrable functions
$\mathbf{R}$	Category of description spaces
Vect	Category of vector spaces
$\mathbf{R}(\mathbf{a})$	$= \mathbf{R}$ in $[\mathbf{a}^1, \ldots, \mathbf{a}^m]$

I. INTRODUCTION

> "Where shall I begin, please your Majesty?" he asked. "Begin at the beginning," the King said, very gravely, "and go on till you come to the end: then stop."
>
> Lewis Carroll, *Alice in Wonderland*

The single greatest influence in the development of this work is the book "Fundamentals of Measurement and Representation of Natural Systems" by Professor Robert Rosen (1978, Elsevier North-Holland, Inc., New York).

I have always been more interested in "conceptual" biomathematics than in looking at specific models of a biological process. In other words, my interest is oriented more toward the *logic* of mathematical biology. During the course of many stimulating discussions with Professor Rosen, we talked about the theories of measurement, recognition, discrimination, interactions, bifurcations, and classifications, and above all, he taught me the importance of the idea of *alternate descriptions* (Rosen, 1976) of a system. After having learned to think plurally and after having read the book (Rosen, 1978), it became quite natural to me to work on the problems of measurement, interaction, and representation from the standpoint of the theory of categories, which is the logical mathematical tool to be used when one studies a collection of objects with similar structures.

In his book *The Scientific Outlook* (1931) Bertrand Russell described the "scientific process" as composed of three main stages; the first consists in observing the significant facts; the second in arriving at a hypothesis which, if it is true, would account for these facts; the third in deducing from this hypothesis further consequences which can be tested by observation. So according to Russell, the act of *observation* is basic to science. Intuitively, the notion of *observable* is attached to that of a concrete procedure for determining the value assumed by the observable of a system at a specific time. The crucial ingredient of any such procedure is a measuring instrument, which forms the basis both for our knowledge of the physical world and for our formulation of models that organize this knowledge and allow us to predict and control.

Rosen (1978) provides a comprehensive theory of observables and the descriptions arising from them. The theory is then applicable to any situation in which objects of interest are labelled by definite mapping processes, measurement in physics, pattern recognition, discrimination, or classification. All these diverse situations share a common character, namely, the generation of numbers (or other kinds of invariants) that serve to label the processes with which they are associated, such that processes are considered "the same" if and only if they bear the same label. This leads to the idea of observable induced equivalence relations.

It is suggested in Rosen (1978) that a formal treatment of systems with observables using category theory would be a fruitful undertaking. The engagement in this problem marked the beginning of my dissertation. The next idea immediately came to mind: Why stop at systems with observables? Since the process of observation ultimately rests on the capacity of a given system to induce a dynamics (i.e., a change of state) in a measuring instrument (alias meter, recognizer, discriminator, classifier, etc.), it seems natural to consider systems with dynamics as well. My definition of *dynamics* is more general than that in Rosen. Roughly, the trajectories of my dynamics are not defined on the whole real line as in Rosen, but are only required to be defined on a (possibly bounded) neighbourhood (a, b) of zero, in which case the "observed value" of the dynamical interaction can simply be defined as the value at b, whereas in the former case one needs asymptotic values.

There is a reciprocity that exists in general dynamical interactions between systems. The process of measurement can be considered as a reciprocal induction of dynamics in both the system being measured and the system that measures. Then the basic problem in the analysis becomes to determine the observables through which a particular given dynamics is taking place, to specify the subsystems to which these observables belong, and to identify the manner in which each of these subsystems is causing the others to change states.

The concepts of linkage of observables, stability, and bifurcation, and their connections with dynamics are introduced in Rosen (1978). In my study I treated them in the context of the category of natural systems, an amalgamation of states, observables, and dynamics. Thus my formal treatment of natural systems was initiated by, but very different from, Rosen's book. Further, the categories I constructed have many curious links with diverse branches of mathematics from topology to Galois theory and provide a natural setting to discuss the modelling relation.

During the course of this work, some questions suggest themselves naturally: the problems of growth, differentiation, development, and aging arose when I considered the evolution of natural systems through time. These problems converged to the concept of the organism, which led to the organismic sets of Nicolas Rashevsky (1972) and the living systems of James Miller (1978). At the time of writing this thesis, I was also collaborating with Professor I. W. Richardson on a paper (Richardson, Louie, and Swaminathan, 1982) in which we developed a phenomenological calculus underlying his theory of the description space. I was pleasantly surprised to find a connection between the ideas of categorical system theory in my thesis and this phenomenological calculus. It is quite remarkable that the apparently completely different methods used by my two teachers, Professors Rosen and Richardson, to analyze complex natural systems are actually very close in essence through this connection woven into my thesis.

From whatever side we approach our principle, we reach the same conclusion.

II. PRELIMINARIES: CATEGORY THEORY

> There is therefore only a single categorical imperative and it is this: Act only on that maxim through which you can at the same time will that it should become a universal law.
>
> Immanuel Kant

1. General Discussion

Many problems in mathematics are not primarily concerned with a single object such as a function, a group, or a measure, but deal instead with large classes of such objects. The classes consist of sets with a given structure and of the mappings preserving this structure.

Thus we may be dealing with groups; the mappings preserving the group structure are the (group-) homomorphisms. When we turn to vector spaces, the appropriate mappings are the linear transformations. And when we study topological spaces, continuous functions arise naturally.

A useful discussion of this situation can be given within a general framework that assumes only very little about the mappings. All we need is that they are closed under composition and include the identity mapping. But we need not even assume them to be mappings. The definition takes the following form.

2. *Category*

DEFINITION. A *category* **C** consists of

(1) a collection of *objects*.

(2) For each pair of **C**-objects A, B, a set $\mathbf{C}(A, B)$, the *hom-set* of *morphisms* from A to B. If $f \in \mathbf{C}(A, B)$, we also write $f: A \to B$.

(3) For any three objects A, B, C a mapping

$$\mathbf{C}(A, B) \times \mathbf{C}(B, C) \to \mathbf{C}(A, C)$$

taking $f: A \to B$ and $g: B \to C$ to its *composite* $g \circ f: A \to C$.

These satisfy the following three axioms:

(i) $\mathbf{C}(A, B) \cap \mathbf{C}(C, D) = \varnothing$ unless $A = C$ and $B = D$. (Thus each morphism $f: A \to B$ determines uniquely its *domain* $A = \operatorname{dom} f$ and its *codomain* B. So the objects in a category **C** are really redundant and one can simply consider **C** as a collection of morphisms, or *arrows*.)

(ii) *Associativity:* If $f: A \to B, g: B \to C, h: C \to D$, (so that both $h \circ (g \circ f)$ and $(h \circ g) \circ f$ are defined), then $h \circ (g \circ f) = (h \circ g) \circ f$.

(iii) *Identity:* For each object A, there exists $1_A: A \to A$ such that for any $f: A \to B$, $g: C \to A$, one has $f \circ 1_A = f$, $1_A \circ g = g$. (It is clear that 1_A is unique.) 1_A is called the *identity morphism* on A.

3. *Isomorphisms*

Speaking informally, two mathematical systems of the same nature are said to be *isomorphic* if there is a one-to-one mapping of one onto the other that preserves all relevant properties, or a "structure-preserving bijection." Such a mapping is an *isomorphism* and it usually coincides with the intuitively most natural concept of structural preservation. Categorically, we have the

DEFINITION. A morphism $f: A \to B$ is an *isomorphism* if there exists an *inverse morphism* $g: B \to A$ such that $g \circ f = 1_A$ and $f \circ g = 1_B$. It is clear that if such inverse morphism exists it is unique and may be denoted by f^{-1}.

An isomorphism with domain and codomain both A is an *automorphism* on A. If there exists an isomorphism from A to B then A and B are *isomorphic*, denoted by $A \cong B$. Isomorphic objects are abstractly the same and most constructions of category theory are only unique "up to isomorphism."

4. Examples of Categories

(1) The category **Ens** of sets and functions (i.e., the class of **Ens**-objects is the class of all sets). (In this study we shall take the naive viewpoint of set theory and assume the existence of a suitable universe of all sets—all *small* sets.) And for sets A and B, $\mathbf{Ens}(A, B) = B^A$ is the set of all functions from A to B. Note that $f: A \to B$ and $f: A \to B'$, where $f(A) \subset B$, $f(A) \subset B'$, and $B \neq B'$, are considered as different **Ens**-morphisms even though as mappings they are the same. It is understood that similar remarks apply to the other examples that follow. Also, **Ens**-isomorphisms are just bijections.

(2) The category **Gp** of groups and homomorphisms.

(3) The category **Top** of topological spaces and continuous functions. **Top**-isomorphisms are homeomorphisms.

(4) The category of nonempty sets and functions.

(5) Let M be a monoid. We can regard M as a category with a single object, whose morphisms are the elements of M. Note the identity in the monoid is the identity morphism and composition of morphisms is the monoid operation. If all morphisms are isomorphisms, the monoid is a group, and conversely. (Indeed, if A is any object in a category $\mathbf{C}$, the hom-set $\mathbf{C}(A, A)$ can be considered as a monoid.)

(6) A partially ordered set $(S, \leq)$ may be considered as a category whose objects are the elements of S, and $S(a, b)$ for $a, b \in S$ has a single element or is empty according to whether $a \leq b$ or not.

From these examples we see that morphisms are generally, but not always, mappings. We shall, most of the time, illustrate the composition of morphisms by diagrams as for mappings. For example, the associativity condition of II.2(ii) is equivalent to saying that in

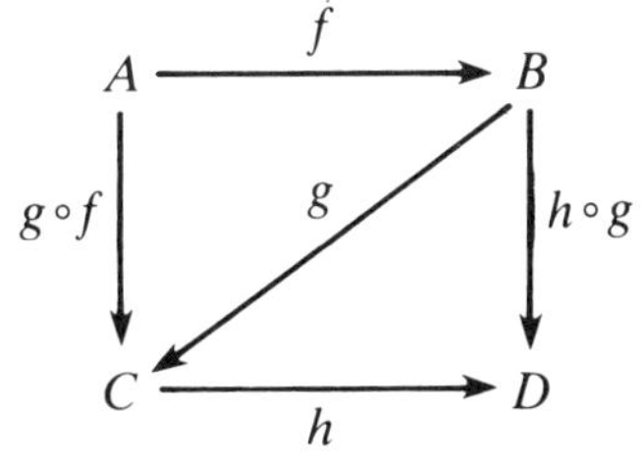

the square commutes.

5. Subcategory

Given categories **A** and **B**, we say that **A** is a *subcategory* of **B** if each **A**-object is a **B**-object, each **A**-morphism is a **B**-morphism, and composition of morphisms is the same in **A** and **B**. Thus for any two **A**-objects X and Y, we have

$$\mathbf{A}(X, Y) \subset \mathbf{B}(X, Y).$$

If equality holds for all X and Y, **A** is a *full subcategory* of **B**.

The category of sets and injections is a subcategory of **Ens** that is not full, while the category of nonempty sets and functions is a full subcategory of **Ens**. The category **Gp** is a full subcategory of the category **Mon** of monoids and monoid-homomorphisms—every monoid-homomorphism between groups is actually a group-homomorphism; and the category **Ab** of abelian groups is a full subcategory of **Gp**.

Note that for a general subcategory we must specify both objects and morphisms and for a full subcategory we need only specify the objects.

6. *Functors*

The idea of a "structure-preserving mapping" can be extended to categories themselves. Let **A** and **B** be categories; then a (*covariant*) *functor* from **A** to **B** is a function $F: \mathbf{A} \to \mathbf{B}$ that assigns to each **A**-object X a **B**-object FX and to each **A**-morphism $f: X \to Y$ a **B**-morphism $Ff: FX \to FY$ such that

(i) If $g \circ f$ is defined in **A**, then $Fg \circ Ff$ is defined in **B** and $F(g \circ f) = Fg \circ Ff$.

(ii) For each **A**-object X, $F1_X = 1_{FX}$.

Besides the covariant functors there is another kind of functor, which reverses the composition. A *contravariant functor* F from **A** to **B** assigns to each **A**-object X a **B**-object FX and to each **A**-morphism $f: X \to Y$ a **B**-morphism $Ff: FY \to FX$ such that (ii) above holds and

(i*) If $g \circ f$ is defined in **A**, then $Ff \circ Fg$ is defined in **B** and $F(g \circ f) = Ff \circ Fg$.

7. *Dual Category*

Associated with each category **C** there is another category called its *dual* or *opposite*, denoted by $\mathbf{C}^{\text{op}}$, formed by "reversing all the arrows." Explicitly, $\mathbf{C}^{\text{op}}$ has the same objects as **C**, but to each **C**-morphism $f: X \to Y$ there corresponds a $\mathbf{C}^{\text{op}}$-morphism $f^{\text{op}}: Y \to X$, so that $f^{\text{op}} \circ g^{\text{op}}$ is defined whenever $g \circ f$ is defined, and that $(g \circ f)^{\text{op}} = f^{\text{op}} \circ g^{\text{op}}$.

Note that a contravariant functor from **A** to **B** may be described as a covariant functor from $\mathbf{A}^{\text{op}}$ to **B**, or as a covariant functor from **A** to $\mathbf{B}^{\text{op}}$. Also, $(\mathbf{A}^{\text{op}})^{\text{op}} = \mathbf{A}$.

8. Examples of Functors

(1) There is a covariant functor from **Gp** (or **Top**, or any category of "sets with structure" in general) to **Ens** which assigns to each group (or topological space) its underlying set, and regards each homomorphism (or continuous function) as a mapping of sets. This functor is called the *forgetful functor*.

(2) If **A** is a subcategory of **B**, then the inclusion of **A** in **B** is a functor, the *inclusion functor*.

(3) Let M and N be monoids, regarded as categories [see Example II.4(5)]. A covariant functor from M to N is simply a monoid-homomorphism.

(4) For any category **C** and **C**-object A, there is a covariant functor h^A [and a contravariant functor h_A] from **C** to **Ens** that assigns to a **C**-object X the set $h^A X = \mathbf{C}(A, X)$ [and $h_A X = \mathbf{C}(X, A)$] and to a **C**-morphism $f: X \to Y$ the function $h^A f: \mathbf{C}(A, X) \to \mathbf{C}(A, Y)$ [and $h_A f: \mathbf{C}(Y, A) \to \mathbf{C}(X, A)$] defined by

$$h^A f(g) = f \circ g \qquad \text{for} \quad g: A \to X$$

$$[\text{and } h_A f(g) = g \circ f \qquad \text{for} \quad g: Y \to A],$$

i.e., via the diagrams

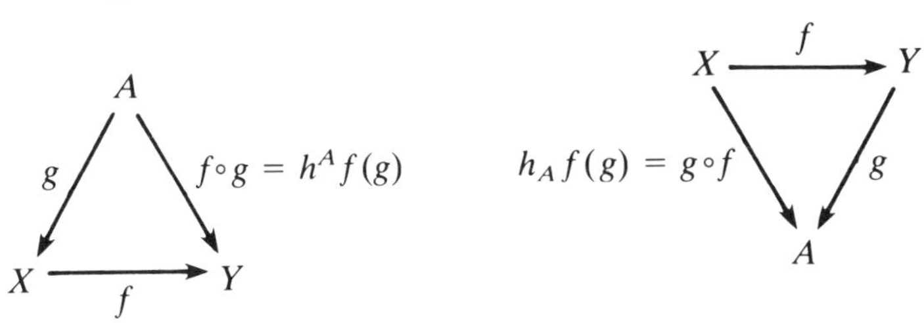

h^A and h_A are known as the *covariant hom-functor* and the *contravariant hom-functor*, respectively.

(5) The *homology functors* H_n from **Top** to **Ab** that take a topological space X to its singular homology groups $H_n(X)$, and a continuous function $f: X \to Y$ to the homomorphisms $f_*: H_n(X) \to H_n(Y)$ are covariant functors. Similarly, the *cohomology functors* H^n: **Top** $\to$ **Ab** are contravariant functors. In fact, it is in the study of algebraic topology that the ideas of category theory originated (see Eilenberg and Mac Lane, 1945). It is interesting to note that category theory arises from the area of mathematics that gives an example of the power of the modelling relation. (In algebraic topology the topological spaces can be considered as being "modeled" by various algebraic objects. For a discussion of the modelling relation, see Section VI.)

THEOREM. Let $F: \mathbf{A} \to \mathbf{B}$ be a (covariant or contravariant) functor. Then F maps **A**-isomorphisms to **B**-isomorphisms. □

Applied to algebraic topology, for example, this theorem says that if $H_n(X)$ and $H_n(Y)$ are not isomorphic groups, then the topological spaces X and Y are not homeomorphic. This is why algebraic topology is used as a tool in dealing with the "homeomorphism problem."

9. The Category **Cat**

The idea of category can be applied to categories and functors themselves. Functors can be composed—given functors $F: \mathbf{A} \to \mathbf{B}$ and $G: \mathbf{B} \to \mathbf{C}$ the maps $X \mapsto G(FX)$ and $f \mapsto G(Ff)$ on $\mathbf{A}$-objects X and $\mathbf{A}$-morphisms f define a functor $G \circ F: \mathbf{A} \to \mathbf{C}$. This composition is clearly associative. For each category $\mathbf{A}$ there is an *identity functor* $I_{\mathbf{A}}: \mathbf{A} \to \mathbf{A}$. So we may consider the category **Cat** having as objects all *small* categories and as morphisms all functors between them. The reason we have to consider small categories and not any category in general is that we want the **Cat**-objects to be in the universe of set theory (see Example II.4(1)) and hence we want the collection of objects of a **Cat**-object to be a *small set*.

An *isomorphism* $F: \mathbf{A} \to \mathbf{B}$ of categories is a functor that is a bijection both on objects and on morphisms. This is clearly equivalent to the existence of an "inverse functor" $F^{-1}: \mathbf{B} \to \mathbf{A}$.

10. Full and Faithful Functors

A functor $F: \mathbf{A} \to \mathbf{B}$ is *full* if to each pair X, Y of $\mathbf{A}$-objects and to every $\mathbf{B}$-morphism $g: FX \to FY$ there is an $\mathbf{A}$-morphism $f: X \to Y$ such that $g = Ff$.

A functor $F: \mathbf{A} \to \mathbf{B}$ is *faithful* if to each pair X, Y of $\mathbf{A}$-objects and to every pair $f_1, f_2: X \to Y$ of $\mathbf{A}$-morphisms the equality $Ff_1 = Ff_2: FX \to FY$ implies $f_1 = f_2$.

For example, if $F: M \to N$ is a surjective but not injective monoid homomorphism, then when M and N are regarded as categories [Example II.4(5)], F can be considered as a functor and it is full but not faithful. On the other hand, the forgetful functor from **Gp** to **Ens** is faithful but not full.

These two properties may be visualized in terms of hom-sets. For each pair of $\mathbf{A}$-objects X and Y, the functor $F: \mathbf{A} \to \mathbf{B}$ assigns to each $f \in \mathbf{A}(X, Y)$ a morphism $Ff \in \mathbf{B}(FX, FY)$, and so defines a function

$$F_{X,Y}: \mathbf{A}(X, Y) \to \mathbf{B}(FX, FY)$$

with $F_{X,Y}(f) = Ff$. Then F is full when every such function is surjective and faithful when every such function is injective. If F is both full and faithful, then every such function is a bijection, but this does not mean that F is a **Cat** isomorphism, for there may be $\mathbf{B}$-objects that are not in the image of F.

If **A** is a subcategory of **B**, then the inclusion functor is faithful. It is full if and only if **A** is a full subcategory of **B**.

As another example, consider Rosen's (1958) representation theorem of biological systems: "Given any system M and a resolution of M into components, it is possible to find an abstract block diagram which represents M and which consists of a collection of suitable objects and mappings from the category of all sets," i.e., there is a faithful functor from the "category of all systems" to **Ens**.

11. Natural Transformation

DEFINITION. Suppose $F, G: \mathbf{A} \to \mathbf{B}$ are two functors between the same two categories. A *natural transformation* α from F to G is defined by

(1) for each **A**-object X there is a **B**-morphism $\alpha X: FX \to GX$,

(2) for each **A**-morphism $f: X \to Y$ the following square of **B**-morphisms commutes:

$$\begin{array}{ccc} FX & \xrightarrow{\;Ff\;} & FY \\ {\scriptstyle \alpha X}\downarrow & & \downarrow{\scriptstyle \alpha Y} \\ GX & \xrightarrow[\;Gf\;]{} & GY \end{array}$$

We use the notation $\alpha: F \to G$ when α is a natural transformation from F to G.

12. Functor Category

If **A** and **B** are categories, we can define the *functor category* $\mathbf{B}^{\mathbf{A}}$ to have as objects all (covariant) functors from **A** to **B**, to have as morphisms natural transformations, and to have as composition and identities the "pointwise" ones. It is easy to check that $\mathbf{B}^{\mathbf{A}}$ is indeed a category.

Isomorphisms in $\mathbf{B}^{\mathbf{A}}$ are called *natural equivalences*. The natural transformation $\alpha: F \to G$ is a natural equivalence if and only if each αX (for **A**-objects X) is an isomorphism (from FX to GX) in **B**.

13. Products

In set theory, the product of a family $\{A_i : i \in I\}$ of sets is the set $A = \Pi A_i$ of all I-tuples $(a_i : i \in I)$ with each $a_i \in A_i$. The function $\pi_j: A \to A_j$ that sends (a_i) to a_j is called the *jth projection* (or projection onto the jth coordinate). If A' is a set and if there are functions $f_j: A' \to A_j$, then there exists a unique function $f: A' \to A$ such that $\pi_j \circ f = f_j$ for all $j \in I$; namely, $f(a') = (f_i(a') : i \in I)$.

With the preceding motivation, given a family of objects $\{A_i : i \in I\}$ in a category **C**, a *product* of this family in **C** is an object A of **C**, often denoted by ΠA_i, with an I-tuple of **C**-morphisms (*projections*) $\pi_i : A \to A_i$, possessing the *universal property* that whenever A' is a **C**-object similarly equipped with $f_i : A' \to A_i$, there exists a unique **C**-morphism $f: A' \to A$ such that for all $i \in I$, $\pi_i \circ f = f_i$:

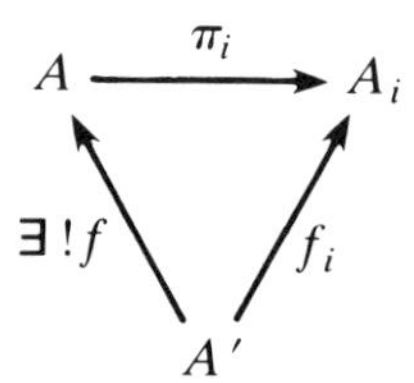

A category **C** *has products* if ΠA_i exists for every family $\{A_i : i \in I\}$.

LEMMA. Any two products of $\{A_i : i \in I\}$ are isomorphic. □

This lemma holds for all universal properties, i.e., any constructions in a category via universal properties are unique up to isomorphism.

DEFINITION. The product of the empty family in **C** is the *final object* of **C**. X is the final object of **C** if and only if for every **C**-object A there is a unique **C**-morphism from A to X.

EXAMPLES. **Ens** has products, the usual "cartesian product sets." The final object in **Ens** is a singleton set. **Top** has products, the cartesian product sets with the product topologies. The final object is a singleton set with the only topology. **Gp** has products, the direct products of groups. The final object is the trivial group.

14. Duality

For each concept in a category **C**, there is a *co-concept* from its dual category $\mathbf{C}^{op}$. For example, if $f: A \to B$ is a **C**-morphism, then A is the *domain* of f in **C**; now $f^{op}: B \to A$ in $\mathbf{C}^{op}$ is such that B is the domain of f^{op}, hence B is the *codomain* of f in **C**. This is consistent with Definition II.2(i).

If $\Sigma(\mathbf{C})$ is a statement about an arbitrary category **C**, let Σ^{op} be the statement defined by $\Sigma^{op}(\mathbf{C}) = \Sigma(\mathbf{C}^{op})$. For example, consider the statement in Lemma II.13:

$$\Sigma(\mathbf{C}) = \text{Products in a category } \mathbf{C} \text{ are unique up to isomorphism.}$$

Then

$$\Sigma(\mathbf{C}^{op}) = \text{Products in a category } \mathbf{C}^{op} \text{ are unique up to isomorphism.}$$

Hence we have the Lemma II.13^{op}:

$\Sigma^{\text{op}}(\mathbf{C}) = \Sigma(\mathbf{C}^{\text{op}})$ = *Coproducts* in a category $\mathbf{C}$ are unique up to isomorphism.

(Note that co-isomorphisms are isomorphisms.)

The Principle of Categorical Duality is: Σ^{op} is *universally* true if Σ is. (Note that "universally true" means that the statement is a consequence of the category axioms.) Duality cuts the work in half.

15. Coproducts

A $\mathbf{C}$-object A with $\mathbf{C}$-morphisms $\iota_j: A_j \to A$ is the *coproduct* of the family $\{A_i : i \in I\}$ in $\mathbf{C}$ if, of course, $(A, \iota_j^{\text{op}}: A \to A_j)$ is the product of $\{A_i : i \in I\}$ in $\mathbf{C}^{\text{op}}$. Explicitly, (A, ι_j) is the coproduct of $\{A_i\}$ in $\mathbf{C}$ if for $(A', g_j: A_j \to A')$ there exists a unique $\mathbf{C}$-morphism $g: A \to A'$ such that for all $j \in I$, $g \circ \iota_j = g_j$. The morphisms ι_j are called *injections* and the coproduct of $\{A_i\}$ is denoted by $\coprod A_i$.

Diagrammatically, we have:

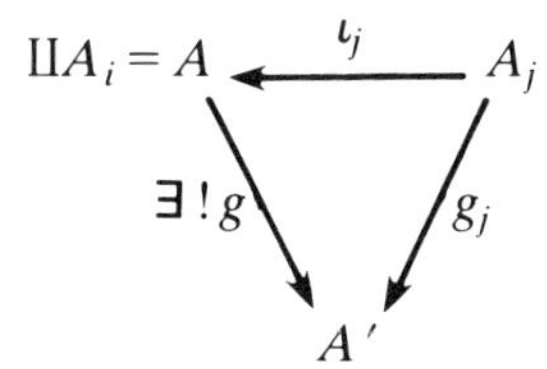

A category $\mathbf{C}$ *has coproducts* if $\coprod A_i$ exists for every family $\{A_i : i \in I\}$.

The coproduct of the empty family is the *initial object* of $\mathbf{C}$. X is the initial object of $\mathbf{C}$ if and only if for every $\mathbf{C}$-object A there is a unique $\mathbf{C}$-morphism from X to A.

EXAMPLES. **Ens** has coproducts, the disjoint union $\coprod A_i = \Pi\{i\} \times A_i = \{(i, a) : i \in I, a \in A_i\}$ and the jth injection ι_j sending $a \in A_j$ to (j, a). The unique initial object of **Ens** is the empty set $\varnothing$. (By the Bourbaki convention, $X^{\varnothing} = \{\varnothing\}$ for any set X; i.e., there is a unique function from the empty set to any set, namely, the inclusion map of the empty set into X—the "empty function.")

Top has coproducts, the disjoint union equipped with the direct sum topology—a set G in $\coprod A_i$ is open if and only if $G \cap A_i$ is open for each i. The initial object is the empty space.

The coproduct in categories of sets with structure may have its underlying set different from the disjoint union. For example, in **Ab**, the coproduct $\coprod A_i$ is the subgroup of ΠA_i consisting of all I-tuples (a_i) such that $a_i = e_i$ = the identity of A_i for all but finitely many i. $\coprod A_i$ is usually referred to as the direct sum $\bigoplus A_i$. And for the universal property, $g: \amalg A_i \to A'$ is defined by $g((\bar{a}_i : i \in I)) = \Sigma g_i(a_i)$. Note that $a_i = e_i$ hence $g_i(a_i) = e' =$ the identity of A' for all but finitely many i, so there is no "convergence" problem in the sum. The initial object of **Ab** is the trivial group that is also the final object. In a category **C**, an object that is both initial and final is called the *zero object*.

This ends the preliminaries on the fundamentals of category theory, which is put here for the notation and for completeness. Many other constructions are possible in a category, and they will be introduced and defined when encountered. The reader is referred to any standard text on category theory (e.g., Mac Lane, 1971) for more detailed discussions.

III. THE CATEGORY OF FORMAL SYSTEMS

> To banish reality is to sink deeper into the real; allegiance to the void implies denial of its voidness.
>
> Seng-Ts'an

A. Introduction

1. General Discussion

Throughout this study we will be dealing with three basic undefined terms: *system*, *state*, and *observable*. Intuitively, a *system* is some part of the real world that is our object of study; a *state* is a specification of what our system is like at a particular time; and an *observable* of the system is some characteristic of the system that can, at least in principle, be measured. In other words, an observable of a system is a quantity that can induce dynamics in some appropriate meter.

These three basic concepts are interrelated via two fundamental propositions that we shall take as axioms in all of what follows. These propositions are as follows:

PROPOSITION 1. The only meaningful physical events that occur in the world are those represented by the evaluation of observables on states.

PROPOSITION 2. Every observable can be regarded as a mapping from states to real numbers.

Scientific activity usually starts with the collection of observable phenomena within a given field. The significance of Proposition 1 lies in the word *meaningful*—because our information on a system is derived from what we can observe; hence what we cannot observe (in the generalized sense of creating percepts in our brains) will not be meaningful to us. This points also to the fact that as our means of observation increase, the more "alternate descriptions" we have on a system, we will have more insights in how and why things work.

To see why Proposition 2 makes sense, let us first quote Sir Arthur Eddington from his *Fundamental Theory* (1949): "The whole subject matter of exact science consists of pointer readings and similar indications; whatever quantity we say we are 'observing', the actual procedure nearly always ends in reading the position of some kind of indicator on a graduated scale or its equivalent." In science the most common form of questioning Nature is through experiments, and the answer often comes in *numerical* readings.

It may well be that from the point of view of scientific enquiry, the only way we can handle any entity is through the numbers associated with its measurement, which in turn is defined through the measuring instrument, the meter. This, in particular, is the view of Einstein's on *time*. Such a definition of time—as something dependent of a measuring instrument—is called an *operational definition*. Borrowing the terminology, our two propositions are then saying that all our observables are operationally defined, and all our meters are real-valued.

It must be realized, however, that the operational definition of scientific entities will enable us to investigate only certain aspects of science. Other questions are in principle unanswerable (i.e., they will not yield a measurement that gives an answer to the question). Such questions are then meaningless in the context of scientific investigation. But the study of those aspects amenable to science based on operational definitions is enough to keep us busy forever. After all, the study of a model of the *whole* of Nature belongs to metaphysics, not science.

2. *Equivalence Relations*

Let us consider the prototype situation, in which we have a set S of states and a real-valued function $f: S \to \mathbb{R}$ that represents an observable. f induces an equivalence relation R_f on S defined by

$$s_1 R_f s_2 \quad \text{iff} \quad f(s_1) = f(s_2).$$

Clearly, the quotient set S/R_f is in one-to-one correspondence with the spectrum $f(S)$. In general the observable f conveys limited information about its domain S, because by definition it cannot distinguish between states lying in the same equivalence class, and the set of states of our system would appear to be S/R_f. This is why "alternate descriptions" of a system are important: The more observables we have, the more information we have on S.

On the other hand, the equivalence relations on S induce an equivalence $\sim$ on $\mathbb{R}^S$, the set of all real-valued functions on S, as follows. For $f, g \in \mathbb{R}^S$, define $f \sim g$ if and only if $R_f = R_g$, i.e., if and only if $f(s_1) = f(s_2)$ is equivalent to $g(s_1) = g(s_2)$. Two equivalent observables "convey the same information" about the elements of S; therefore we cannot distinguish between elements of S by employing equivalent observables. Note, however, that $f \sim g$ only means that $S/R_f = S/R_g$, i.e., there is a one-to-one correspondence between $f(S)$ and $g(S)$, and there need be no relation whatsoever between the values of $f(s)$ and $g(s)$ for $s \in S$. In particular, $|f(s_1) - f(s_2)|$ small does not imply $|g(s_1) - g(s_2)|$ small. Thus if we are considering the metric aspects of observables (which we shall do in later sections), we cannot pass to the equivalence classes in $\mathbb{R}^S/\sim$. But when we are only interested in the *reduced states* in S/R_f and not the specific values of $f(S)$, it is more convenient to consider observables as elements of $\mathbb{R}^S/\sim$. One reason for this is that $\mathbb{R}^S/\sim$ is a partially ordered set under the relation *refinement*. [Define f *refines* g, $f \leq g$, if $f(s_1) = f(s_2)$ implies $g(s_1) = g(s_2)$, i.e., if $R_f \subset R_g$, for $f, g \in \mathbb{R}^S$. Then it is clear that $\leq$ is reflexive and transitive on $\mathbb{R}^S$ and hence is a preorder. But $f \leq g$ and $g \leq f$ only implies $R_f = R_g$ (i.e., $f \sim g$) and not $f = g$. So to make $\leq$ antisymmetric, one passes onto $\mathbb{R}^S/\sim$. Note that $f \leq g$ and $g \leq f$ if and only if $f \sim g$.] It is not uncommon in mathematics to consider equivalence classes of functions instead of the functions themselves—the L^p spaces, for example, are equivalence classes of functions with $f \sim g$ if and only if $f = g$ almost everywhere.

B. The Category S

We shall now undertake a formal treatment of systems, considered as abstract mathematical objects. We shall be studying the properties of the category **S** of (formal) systems and comparing them to those of the category **Ens** of sets.

3. *Objects*

An object of **S**, a formal system, shall consist of a pair (S, F), where S is a set and F is a set of real-valued functions on S. The elements of S are the states and the elements of F are the observables of the formal system. We shall always assume $0 \in F$ (where 0 is the zero function on S sending all states to the number 0) although for brevity we may sometimes omit 0 when we list the elements of F in specific examples. Thus we have F nonempty in order to avoid "empty set pathologies." The observable 0 is simply "identifying the states in S." Note that $(0)_\sim$ is the greatest element in the partially ordered set $(\mathbb{R}^S/\sim, \leq)$.

4. Morphisms

An **S**-morphism $\phi \in \mathbf{S}((S_1, F_1), (S_2, F_2))$ is a pair of functions $\phi \in \mathbf{Ens}(S_1, S_2)$ and $\phi \in \mathbf{Ens}(F_1, F_2)$, such that for all $f \in F_1$ for all $s, s' \in S_1$, $f(s) = f(s')$ implies $(\phi f)(\phi s) = (\phi f)(\phi s')$, i.e., $sR_f s'$ implies $(\phi s)R_{\phi f}(\phi s')$.

Note that this "compatibility" condition is equivalent to saying for all $G \subset F_1$ for all $s, s' \in S_1$, $sR_G s'$ implies $(\phi s)R_{\phi G}(\phi s')$, where $R_G = \cap R_f : f \in G$ (hence $sR_G s'$ if and only if for all $f \in G$ $f(s) = f(s')$) and $\phi G = \{\phi f : f \in G\} \subset F_2$. This means that for all $G \subset F_1$, ϕ can be considered as a mapping from S_1/R_G to $S_2/R_{\phi G}$.

We always define $\phi 0 = 0$. This is compatible because clearly $0s = 0s'$ implies $0(\phi s) = 0(\phi s')$. Note also that for any observable f, the assignment $\phi f = 0$ is acceptable.

5. Identity

Define $1_{(S,F)} \in \mathbf{S}(S, F), (S, F))$ by for all $s \in S$ $s \mapsto s$ and for all $f \in F$ $f \mapsto f$. (Thus for all $G \subset F$, $G \mapsto G$.) Then clearly $1_{(S,F)}$ satisfies the compatibility condition.

6. Composition

Define composition of morphisms in **S** as simultaneously the compositions on the states and on the observables; i.e., if $\phi: (S_1, F_1) \to (S_2, F_2)$ and $\psi: (S_2, F_2) \to (S_3, F_3)$, define $\psi \circ \phi: (S_1, F_1) \to (S_3, F_3)$ by for every $s \in S_1$ $\psi \circ \phi(s) = \psi(\phi(s))$ and for every $f \in F_1$ $\psi \circ \phi(f) = \psi(\phi f)$. Note for $f \in F_1$ and $s, s' \in S_1$, $sR_f s'$ implies $(\phi s)R_{\phi f}(\phi s')$, which in turn implies $\psi(\phi s)R_{\psi(\phi f)}\psi(\phi s')$; so $\psi \circ \phi$ satisfies the compatibility condition.

Clearly, composition so defined is associative, and for $\phi: (S_1, F_1) \to (S_2, F_2)$, $1_{(S_2,F_2)} \circ \phi = \phi = \phi \circ 1_{(S_1,F_1)}$.

7. Isomorphisms

If $\phi: (S_1, F_1) \to (S_2, F_2)$ and $\psi: (S_2, F_2) \to (S_1, F_1)$ are such that $\psi \circ \phi = 1_{(S_1,F_1)}$ and $\phi \circ \psi = 1_{(S_2,F_2)}$, then it is easy to see that $\phi: S_1 \to S_2$ and $\phi: F_1 \to F_2$ must be bijections (**Ens**-isomorphisms) and that for $f \in F_1$ and $s, s' \in S_1$, $f(s) = f(s')$ if and only if $(\phi f)(\phi s) = (\phi f)(\phi s')$, i.e., for every $G \subset F_1$ $S_1/R_G = S_2/R_{\phi G}$.

Thus isomorphic systems are abstractly the same in the sense that there is a "dictionary" (one-to-one correspondence) between the states and between the observables inducing the "same" equivalence relations on the states. In particular, if F and G are two sets of observables on S and there is a bijection $\phi: F \to G$ such that for all $f \in F$ $f \sim \phi f$, then the two systems (S, F) and (S, G)

are isomorphic with the **S**-isomorphism $1_S: S \to S, \phi: F \to G$. Since categorical constructions are only unique up to isomorphism, in the category S all constructions (S, F) are only "unique up to $\sim$-equivalent observables" (i.e., one can always replace F by an $\sim$-equivalent set of observables G in the above sense) even when the set of states S is held fixed. This last comment is particularly important for all constructions in **S** below.

C. Constructions in S

8. *Products*

Products in the category **S** do not always exist. For a family $\{(S_i, F_i): i \in I\}$, the product should be $(S, F) = \Pi(S_j, F_j): j \in I$ with an I-tuple of **S**-morphisms of the form $\pi_i: (S, F) \to (S_i, F_i)$. S is defined as the cartesian product $\Pi\, S_j$ of the sets of states. F is defined as the "cartesian product" $\Pi\, F_j$ of the sets of observables interpreted as follows: for the observable $(f_j: j \in I)$ in F, it is a mapping from S to $\mathbb{R}^I$ defined by

$$(f_j: j \in I)(s_j: j \in I) = (f_j(s_j): j \in I).$$

But **S**-objects can only have real-valued observables so the mapping $f = (f_j: j \in I): S \to \mathbb{R}^I$ must be *represented* by an equivalent mapping from S to $\mathbb{R}$ (such that S/R_f remains the same). In other words, we need a one-to-one map from $(\mathbb{R}^I)^S$ to $\mathbb{R}^S$ that preserves the equivalence relations on S induced by the observables, or what is equivalent, an injection from $\mathbb{R}^I$ to $\mathbb{R}$. An injection from $\mathbb{R}^I$ to $\mathbb{R}$ only exists when the cardinality of I is less than that of $\mathbb{R}$; thus **S** only has countable (including finite) products. Note, however, that the **S**-product is independent of the choice of the injection from $\mathbb{R}^I$ to $\mathbb{R}$ because the effect of changing the injection is simply a switch from F to an $\sim$-equivalent set G, and (S, F) and (S, G) are isomorphic (see Section III.7). Different choices of the injection define different **S**-isomorphism class representatives of the product.

The projections are obviously defined by $\pi_i((s_j: j \in I)) = s_i$ and $\pi_i((f_j: j \in I)) = f_i$. And it is easily checked that the π_i's are indeed **S**-morphisms.

To see that (S, F) is a product, consider an **S**-object (X, H) equipped with an I-tuple of **S**-morphisms $\pi_i: (X, H) \to (S_i, F_i)$. Then

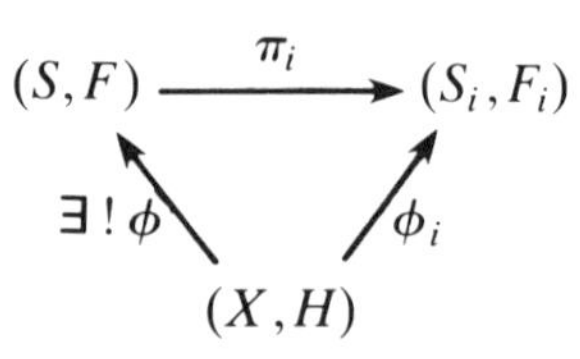

We can define $\phi:(X,H)\to(S,F)$ by for $x\in X$ $\phi(x)=(\phi_j(x):j\in I)$, and for $h\in H$ $\phi(h)=(\phi_j(h):j\in I)$, where again $\phi(h)$ is to be represented by a mapping from S to $\mathbb{R}$ via the injection from $\mathbb{R}^I$ to $\mathbb{R}$. It is clear that ϕ is the unique map that makes the diagram commute. [But ϕ is of course dependent on the choice of the injection from $\mathbb{R}^I$ to $\mathbb{R}$ that determines the product (S,F).] To see that ϕ is an **S**-morphism, let $h\in H$ and $x,x'\in X$ be such that $h(x)=h(x')$. Then for each $i\in I$ $(\phi_i h)(\phi_i x)=(\phi_i h)(\phi_i x')$ because each ϕ_i is an **S**-morphism. Whence by definition $(\phi h)(\phi x)=(\phi h)(\phi x')$. Thus $xR_h x'$ implies $(\phi x)R_{\phi h}(\phi x')$. So (S,F) with the π_i's satisfy the universal property in the definition of a categorical product.

The final object in **S** is $(1,\{0\})$ where 1 is the singleton set, the final **Ens**-object. The unique **S**-morphism from any system to $(1,\{0\})$ is clearly the one that sends all states to 1 and all observables to 0.

9. *An Application: Linkage*

DEFINITION. Let S be a set of states and $f,g\in\mathbb{R}^S$ be observables. Let p_f: $S\to S/R_f$ and $p_g:S\to S/R_g$ be the natural quotient maps. For $(s)_f\in S/R_f$ consider the set of R_g-classes which intersect $(s)_f$, i.e., the set

$$p_g p_f^{-1}(s)_f=\{(s')_g:f(s')=f(s)\}=\{(s')_g:(s')_g\cap(s)_f\neq\varnothing\}.$$

Then we say

(1) g is *totally linked* to f at $(s)_f$ if the preceding set consists of a single R_g-class (necessarily $(s)_g$; i.e., $f(s)=f(s')$ implies $g(s)=g(s')$);

(2) g is *partially linked* to f at $(s)_f$ if this set consists of more than one R_g-class, but is not all of S/R_g;

(3) g is *unlinked* to f at $(s)_f$ if this set is S/R_g.

Also, we say g is *totally linked* to f if it is totally linked at each $(s)_f$ and g is *unlinked* to f if it is unlinked at each $(s)_f$.

From the preceding definition it is immediate that g is totally linked to f if and only if R_f refines R_g, which is equivalent to the existence of an **S**-morphism from $(S,\{f\})$ to $(S,\{g\})$ which sends each $s\in S$ to itself and sends f to g, for the latter statement means precisely that $f(s)=f(s')$ implies $g(s)=g(s')$.

For a set of states S equipped with two distinct observables f and g, there is another equivalence relation on S other than R_f and R_g that is of interest—namely, the intersection $R_{fg}\doteq R_f\cap R_g$. The relation R_{fg} is defined by $sR_{fg}s'$ if and only if $f(s)=f(s')$ *and* $g(s)=g(s')$. Note that there may not be an observable of S that generates the equivalence relation R_{fg}, i.e., although mathematically there exists $h \in \mathbb{R}^S$ such that $R_{fg} = R_h$, the set of all possible observables of S, as a representation of a natural system, may not be all of $\mathbb{R}^S$. Thus R_{fg} is generally a formal construction.

There is always an embedding $\phi: S/R_{fg} \to S/R_f \times S/R_g$ that maps $(s)_{fg} \mapsto ((s)_f, (s)_g)$. Via this embedding, a state $s \in S$ is represented by the pair of numbers $(f(s), g(s))$. This embedding ϕ is in general one to one, but it is onto if and only if f and g are totally unlinked (to each other).

This product representation can be constructed neatly as a categorical product. Consider the two systems $(S, \{f, 0\})$ and $(S, \{g, 0\})$. The **S**-product of these two systems is $(S \times S, F)$, where $F = \{0, (f, 0), (0, g), (f, g)\} \subset \mathbb{R}^{S \times S}$, with the natural projections. Now consider further the system $(S, \{f, g\})$. There exist **S**-morphisms

$$\phi_1: (S, \{f, g\}) \to (S, \{f, 0\})$$
$$\text{and} \quad \phi_2: (S, \{f, g\}) \to (S, \{g, 0\})$$

defined by for every $s \in S$

$$\phi_1(s) = s; \qquad \phi_1 f = f, \quad \phi_1 g = 0$$
$$\text{and} \quad \phi_2(s) = s; \qquad \phi_2 f = 0, \quad \phi_2 g = g.$$

So we have the following diagram:

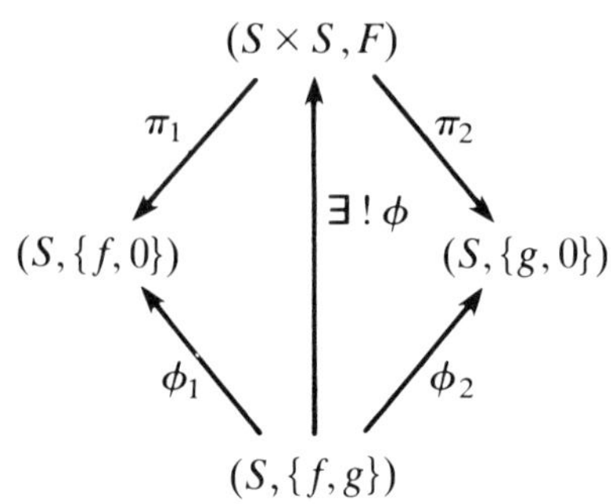

Hence by the universal property of the product, there exists a unique $\phi: (S, \{f, g\}) \to (S \times S, F)$ that makes the diagram commute. Namely, ϕ is defined by sending $s \in S$ to $\phi(s) = (s, s)$—the diagonal map—and by $\phi f = (f, 0)$, $\phi g = (0, g)$. In particular, ϕ being an **S**-morphism implies that $\phi: S/R_{\{f,g\}} \to S \times S/R_{\{(f,0),(0,g)\}}$. It is clear that ϕ is a one-to-one mapping (on S) and that $R_{\{f,g\}} = R_{fg}$ hence $S/R_{\{f,g\}} = S/R_{fg}$. Also, $S \times S/R_{\{(f,0),(0,g)\}} \cong S/R_f \times S/R_g$. Thus ϕ is indeed the one-to-one map from S/R_{fg} to $S/R_f \times S/R_g$, and that the degree of onto-ness of ϕ is an indication of the lack of linkage between f and g. The onto-ness of a morphism is discussed in Section III.14.

10. Equalizers

Let A and B be sets and $f, g: A \to B$ be two functions. The inclusion map i of the subset $E = \{x \in A : f(x) = g(x)\}$ of A can be characterized up to isomorphism by the following universal property: for every function $i': E' \to A$ such that $f \circ i' = g \circ i'$, there exists a unique function $j: E' \to E$ such that $i \circ j = i'$ (since the image of i' is contained in E, j is defined by $j(x) = i'(x)$ for $x \in E'$).

Generalizing this idea, given morphisms f, $g: A \to B$ in a category **C**, an *equalizer* of (f, g) is an object E with a morphism $i: E \to A$ satisfying the following universal property:

(1) $f \circ i = g \circ i$;
(2) given i' with $f \circ i' = g \circ i'$, there exists unique j such that $i \circ j = i'$.

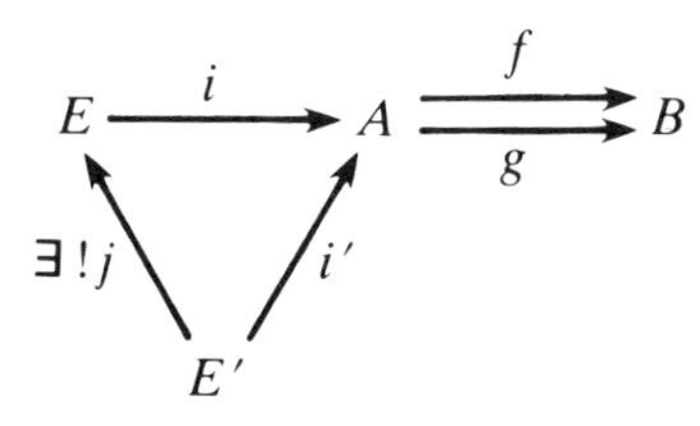

(An equivalent way of defining an equalizer is to say that it is a final object in the category of all **C**-morphisms that satisfy $f \circ i = g \circ i$.) As usual for "universal property" definitions, equalizers are unique up to isomorphism and will be denoted by $\text{eq}(f, g)$. A category **C** *has equalizers* if $\text{eq}(f, g)$ exists for every pair f, $g: A \to B$.

For **S**-morphisms ϕ, $\psi: (S_1, F_1) \to (S_2, F_2)$, $\text{eq}(\phi, \psi) = (E, H)$ may not exist. The equalizer would have to be given by $E = \{s \in S_1 : \phi s = \psi s\}$, $H = \{f|_E : f \in F_1,\ \phi f = \psi f\}$ and $\iota: (E, H) \to (S_1, F_1)$ would be the inclusion morphism. But $\iota(f|_E) = f$ may not be uniquely defined because there may be another $g \in F_1$ such that $g|_E = f|_E$ and $\phi g = \psi g$. Thus an **S**-equalizer only exists when the inclusion from H to F_1 is a single-valued function.

Note when $(E, H) = \text{eq}(\phi, \psi)$ does exist, $\iota: (E, H) \to (S_1, F_1)$ has the property that for all $s, s' \in E$ and for all $g \in H$, $g(s) = g(s')$ if and only if $(\iota g)(s) = (\iota g)(s')$, i.e., $E/R_g \cong \iota(E)/R_{\iota g}$. Further, any **S**-morphism $\chi: (X_1, G_1) \to (X_2, G_2)$ that is one-to-one on the states and observables and that has this property (that $X_1/R_g \cong \chi(X_1)/R_{\chi g}$ for all $g \in G_1$) is an equalizer. It is easy to construct a pair of S-morphisms ϕ_1, ϕ_2 with domain (X_2, G_2) such that $(X_1, G_1) = \text{eq}(\phi_1, \phi_2)$. Thus although **S** does not have equalizers for every pair of **S**-morphisms, given an **S**-morphism ϕ with the correct properties one can always find a pair of **S**-morphisms for which ϕ is the equalizer.

Equalizers will be discussed further in the sections on hierarchies of **S**-morphisms.

11. Coproducts

S has coproducts only when a cardinality condition is satisfied. The coproduct is $(S, F) = \amalg(S_i, F_i)$ where $S = \amalg S_i$ is the coproduct of the S_i's in **Ens** (i.e., the disjoint union $S = \bigcup\{i\} \times S_i)$ and $F = \{0\} \cup \{(i, f) : i \in I, f \in F_i, f \neq 0\}$ defined as follows. For $f \in F_i$, $f \neq 0$, the observable (i, f) of S is defined by

$$(i, f)(j, s) = \begin{cases} f(s) & \text{if} \quad j = i \\ (j, s) \in \mathbb{R} & \text{if} \quad j \neq i \quad (*), \end{cases}$$

where (*) denotes that we assume there is a one-to-one mapping from $\amalg S_j : j \neq i$ to $\mathbb{R}$, i.e., we assume the existence of an observable = on $\amalg S_j : j \neq i$. This assumption is only valid when the cardinality of S is no bigger than that of $\mathbb{R}$. Thus **S**-coproducts only exist when this holds. The natural injections are $\iota_i : (S_i, F_i) \to (S, F)$ with $\iota_i(s) = (i, s)$, $\iota_i f = (i, f)$ for $s \in S_i$ and $f \in F_i$ with $f \neq 0$, and $\iota_i(0) = 0$.

To check the universal property, consider the following:

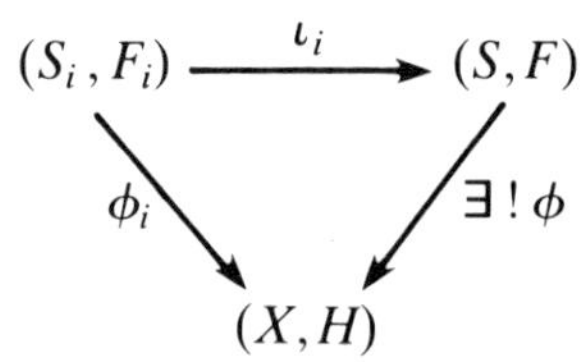

The natural definition for ϕ (given $((X, H), \phi_i)$) is $\phi(i, s) = \phi_i(s)$, $\phi(i, f) = \phi_i f$. Now suppose $(i, f) \in F$, (j, s), $(k, t) \in S$ are such that $(i, f)(j, s) = (i, f)(k, t)$. Then *either* $i = j = k$ and $f(s) = f(t)$ *or* $(j, s) = (k, t)$. If the former, then since ϕ_i is an **S**-morphism, $(\phi_i f)(\phi_i s) = (\phi_i f)(\phi_i t)$, i.e., by definition $(\phi(i, f))(\phi(j, s)) = (\phi(i, f))(\phi(k, t))$. If the latter, then $j = k$ and $s = t$ whence $\phi_j s = \phi_j t$, implying $(\phi_i f)(\phi_j s) = (\phi_i f)(\phi_j t)$, hence $(\phi(i, f))(\phi(j, s)) = (\phi(i, f))(\phi(k, t))$. Thus ϕ is an **S**-morphism.

S, therefore, has an initial object. The initial object in **Ens** is the empty set $\varnothing$, thence the initial object in **S** is $(\varnothing, \{0\})$. For any system (S, F), the unique **S**-morphism from $(\varnothing, \{0\})$ to (S, F) is clearly the empty function on $\varnothing$ with $0 \mapsto 0 \in F$.

12. Coequalizers

$q: B \to Q$ in a category **C** is a coequalizer of $f, g: A \to B$ if, of course, $q^{\text{op}} = \text{eq}(f^{\text{op}}, g^{\text{op}})$ in the dual category $\mathbf{C}^{\text{op}}$.

Ens has coequalizers. Given $f, g: A \to B$ let R be the equivalence relation on B generated by $A' = \{(f(a), g(a)) : a \in A\}$, i.e., let R be the intersection of all equivalence relations on B containing A'. Let $Q = B/R$ with canonical projection $q: B \to Q$. Then $q = \text{coeq}(f, g)$. Clearly, $q \circ f = q \circ g$.

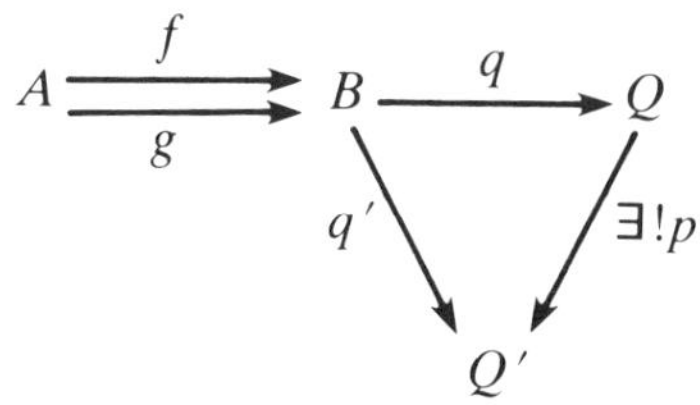

And suppose $q': B \to Q'$ and $q' \circ f = q' \circ g$; then $R' = \{(b_1, b_2): q'(b_1) = q'(b_2)\}$ is an equivalence relation on B containing A' and hence contains R; p is thus defined by $p(b)_R = q'(b)$ and so $p \circ q = q'$.

The category **S** also has coequalizers, constructed as follows. Let ϕ, $\psi: (S_1, F_1) \to (S_2, F_2)$. Let $Q = S_2/R$ where R is the intersection of all equivalence relations on S_2 containing $\{(\phi(s), \psi(s)) \in S_2 \times S_2 : s \in S_1\}$ *and* of all R_{F_2}. So in particular for t, $t' \in S_2$, tRt' implies for all $g \in F_2$ $g(t) = g(t')$. Let $\chi: S_2 \to Q$ be the canonical projection $\chi(t) = (t)_R$. This takes care of the map on the states. As for the observables, let R on F_2 be the intersection of all equivalence relations containing $\{(\phi f, \psi f) \in F_2 \times F_2 : f \in F_1\}$, and let $\chi: F_2 \to H = F_2/R$ be, naturally, $\chi g = (g)_R$, where $(g)_R \in \mathbb{R}^Q$ is to be interpreted as follows. Let $(g)_R: S_2 \to \mathbb{R}$ be such that $R_{(g)_R}$ is the equivalence relation on S_2 generated by $\{R_{g'} : g' \in (g)_R\}$, i.e., $R_{(g)_R}$ is the finest equivalence relation on S_2 such that it is refined by each of the $R_{g'}$, $g' \in (g)_R$. Putting it another way, $R_{(g)R}$ is defined to be the supremum of the family $\{R_{g'} : g' \in (g)_R\}$ in the lattice of all equivalence relations on S_2. It is clear, then, that $R_{(g)_R}$ is refined by R on S_2 and hence $(g)_R$ is well defined on $Q = S_2/R$, so we can consider $(g)_R: Q \to \mathbb{R}$. Finally, to check $\chi: (S_2, F_2) \to (Q, H)$ such defined is indeed an **S**-morphism, let $g \in F_2$ and t, $t' \in S_2$; then $g(t) = g(t')$ implies $(g)_R(t) = (g)_R(t')$ hence $(g)_R(t)_R = (g)_R(t')_R$. So $tR_g t'$ does imply $\chi(t) R_{\chi g} \chi(t')$. And clearly $\chi \circ \phi = \chi \circ \psi$.

Now if $\chi': (S_2, F_2) \to (Q', H')$ is such that $\chi' \circ \phi = \chi' \circ \psi$, then $\{(t, t') \in S_2 \times S_2 : \chi'(t) = \chi'(t')\}$ is an equivalence relation on S_2 containing R. Thus $\pi(t)_R = \chi'(t)$ is well defined on $Q = S_2/R$. Similarly, $\pi(g)_R = \chi' g$ is well defined on H. Clearly, $\chi' = \pi \circ \chi$ and π is unique.

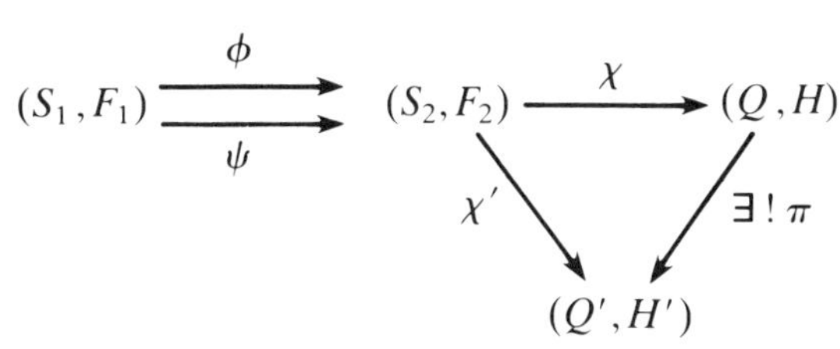

Finally, we have to check that π is an S-morphism. Note that for every $g' \in (g)_R$ (i.e., $g'Rg$), $\chi'g' = \chi'g$ because $\{(g,g') \in F_2 \times F_2 : \chi'g = \chi'g'\}$ is an equivalence relation on F_2 and since for every $f \in F_1$ $\chi'(\phi f) = \chi'(\psi f)$, this equivalence relation contains all $(\phi f, \psi f)$ and hence contains R. Also since χ' is an S-morphism, for each $g' \in (g)_R$ we have $g'(t) = g'(t')$ implying $(\chi'g')(\chi't) = (\chi'g')(\chi't')$, i.e., $(\chi'g)(\chi't) = (\chi'g)(\chi't')$. Thus $R_{\chi'g}$ is "refined" by each of $R_{g'}$ (on F_2). Since $R_{(g)_R}$ is the supremum of $\{R_{g'} : g' \in (g)_R\}$, we have $R_{(g)_R} \subset R_{\chi'g}$. Therefore $(g)_R(t)_R = (g)_R(t')_R$ in Q implies $(g)_R(t) = (g)_R(t')$ in S_2, which in turn implies that $(\chi'g)(\chi't) = (\chi'g)(\chi't')$ in Q', i.e., $\pi(g)_R[\pi(t)_R] = \pi(g)_R[\pi(t')_R]$ in Q', whence $\pi: (Q, H) \to (Q', H')$ is indeed an S-morphism.

It should be obvious that whereas products and equalizers are easy to define in **S**, their dual concepts are a lot more complicated. This is observed in many familiar categories. A difficult problem in the study of a specific category is to describe its coproducts and coequalizers ("colimits") explicitly.

D. Hierarchy of S-Morphisms

13. Monomorphisms

Injective is an important property of functions. In particular it is the tool with which one defines *subsets* and the ordering of cardinals. As it turns out, there are many categorical definitions that characterize injective in **Ens**; we will content ourselves with three of them.

Let $f: A \to B$ in a category **C**. The morphism f is *split mono* if there exists a $g: B \to A$ with $g \circ f = 1_A$, and *equalizer* if $f = \text{eq}(g_1, g_2)$ for some pair g_1, g_2: $B \to C$, and *mono* if for all pairs $g_1, g_2: X \to A$ such that $f \circ g_1 = f \circ g_2$, we have $g_1 = g_2$.

It is a general theorem in category theory (the "hierarchy theorem for monomorphisms;" see Manes, 1976) that split monos are equalizers and equalizers are monos. Also, if f and g are mono or split mono, so is $g \circ f$ (when composition is appropriate); if $g \circ f$ is mono or split mono, so is f. (The analogue for equalizers is not always true; it is true, however, when a category has epi-equalizer factorizations. See Section III.16.)

In **Ens**, monos are the same as injective functions, and it is easy to see that all monos are equalizers. If $f: A \to B$ is mono and A is nonempty, then f is split mono (let $g = f^{-1}$ on $f(A)$ and arbitrary elsewhere). Note, however, that the inclusion map of the empty set into a nonempty set is mono, but never split mono (because $B^{\varnothing} = \{\varnothing\}$ for sets B but $\varnothing^B = \varnothing$ for $B \neq \varnothing$). An **Ens**-mono f from A to B will be denoted $f: A \rightarrowtail B$. This notation will sometimes be borrowed for a monomorphism in any category.

What is a mono in the category **S** like? We claim that it is the same as an **S**-morphism that is injective as set mappings on the set of states and on the set of observables. For suppose $\phi:(S_1,F_1)\to(S_2,F_2)$ is a mono and there are distinct states s and s' in S for which $\phi(s)\neq\phi(s')$, then consider ψ_1, ψ_2: $(S_1,F_1)\to(S_1,F_1)$ with ψ_1 mapping all states in S_1 to s, ψ_2 mapping all states in S_1 to s', and both ψ_1 and ψ_2 acting as identity on F_1. It is easy to check that in this case the **S**-morphisms ψ_1 and ψ_2 are such that $\phi\circ\psi_1=\phi\circ\psi_2$ but $\psi_1\neq\psi_2$, a contradiction. So $\phi: S_1\to S_2$ must be injective. Also, suppose distinct observables f and f' in F_1 are such that $\phi f=\phi f'$, then consider ψ_1, ψ_2: $(\{s\},\{f,f'\})\to(S_1,F_1)$ where $s\in S_1$, ψ_1 is the inclusion, and $\psi_2(s)=s$, $\psi_2 f=f'$, $\psi_2 f'=f$. Again ψ_1, ψ_2 are **S**-morphisms with $\phi\circ\psi_1=\phi\circ\psi_2$ but $\psi_1\neq\psi_2$, a contradiction. So $\phi: F_1\to F_2$ is also injective. Conversely, it is clear that if an **S**-morphism $\phi:(S_1,F_1)\to(S_2,F_2)$ is injective on both S_1 and F_1, it is mono.

Now suppose $\phi:(S_1,F_1)\to(S_2,F_2)$ is an equalizer and that S_1 is nonempty. (F_1 is already nonempty because $0\in F_1$.) Say $\phi=\text{eq}(\psi_1,\psi_2)$ for ψ_1, $\psi_2:(S_2,F_2)\to(X,H)$. Then, as an equalizer, $\phi:(S_1,F_1)\to(S_2,F_2)$ is isomorphic to an inclusion (see Section III.10). So in particular for $f\in F_1$ and $s, s'\in S_1$, $f(s)=f(s')$ if and only if $(\phi f)(\phi s)=(\phi f)(\phi s')$, i.e., $S_1/R_f\cong\phi(S_1)/R_{\phi f}$. Thus ϕ^{-1} is well defined on $\phi(S_1)$ and $\phi(F_1)$ and can be extended to an **S**-morphism on (S_2,F_2). (We need a nonempty S_1 for the same reason as in **Ens**.) So in **S**, an equalizer with nonempty domain is split mono.

In the examples of Section III.15 we will show that a mono is not necessarily an equalizer, so the hierarchy for monomorphisms in **S** is (for $\phi:(S_1,F_1)\to(S_2,F_2)$ with nonempty S_1)

$$\text{split mono}\Leftrightarrow\text{equalizer}\Rightarrow\text{mono}\Leftrightarrow\text{injection (on both } S_1 \text{ and } F_1).$$

14. Epimorphisms

The dual concepts to split mono, equalizer, and mono are, respectively *split epi*, *coequalizer*, and *epi*. Explicitly, $f: A\to B$ (in a category **C**) is *split epi* if there exists a $g: B\to A$ such that $f\circ g=1_B$; f is a *coequalizer* if $f=\text{coeq}(g_1,g_2)$ for some pair $g_1, g_2: C\to A$; f is *epi* if for all pairs $g_1, g_2: B\to X$, $g_1\circ f=g_2\circ f$ implies $g_1=g_2$.

Dually, the hierarchy theorem for epimorphisms states that split epis are coequalizers and coequalizers are epis. Also, if f and g are epi or split epi, so is $g\circ f$; if $g\circ f$ is epi or split epi, so is g. (Again, the analogue for coequalizers is not always true, but is true when the category has coequalizer-mono factorizations; see Section III.15.)

In **Ens**, all epis split and all three concepts mean "surjective": since any function $f: A\to B$ composes equally with χ_B and $\chi_{f(A)}\in 2^B$, it is clear that epis are surjective. The axiom of choice says that surjections are split epi. An **Ens**-

epi (and sometimes a **C**-epimorphism) will be denoted $f: A \twoheadrightarrow B$.

In **S**, an epi is the same as an **S**-morphism that is surjective on both the set of states and the set of observables. For suppose $\phi:(S_1, F_1) \to (S_2, F_2)$ is an epi and there is an $s \in S_2 \sim \phi(S_1)$, then $\psi_1, \psi_2:(S_1, F_1) \to (\{0, 1\}, \{0\})$, where $\psi_1 = \chi_{\phi(S_1)}$ on S_2, $\psi_1 f = 0$ for all $f \in F_2$, $\psi_2 = \chi_{S_2}$ on S_2, $\psi_2 f = 0$ for all $f \in F_2$, provide a pair of **S**-morphisms such that $\psi_1 \circ \phi = \psi_2 \circ \phi$ but $\psi_1 \neq \psi_2$, a contradiction. So $\phi(S_1) = S_2$. Now suppose there is an $f \in F_2 \sim \phi(F_1)$, then $\psi_1, \psi_2:(S_2, F_2) \to (S_2, F_2)$, where $\psi_1 = 1_{(S_2, F_2)}$, $\psi_2 = 1_{S_2}$ on S_2 and $\psi_2 f = 0$ for all $f \in F_2$, is an example in which $\psi_1 \circ \phi = \psi_2 \circ \phi$ but $\psi_1 \neq \psi_2$, again a contradiction. Thus $\phi(F_1) = F_2$. Conversely, it is clear that an **S**-morphism $\phi:(S_1, F_1) \to (S_2, F_2)$ that is onto both S_2 and F_2 is epi.

Thus in **S**, we have

split epi $\Rightarrow$ coequalizer $\Rightarrow$ epi $\Leftrightarrow$ surjection (onto both S_2 and F_2).

In the next section we show that the two preceding one-way implications are indeed irreversible; so the preceding is the hierarchy for epimorphisms in **S**.

Note that although the two hierarchies in **S** for the dual concepts of monomorphisms and epimorphisms are not the same, this is *not* a counter-example to the principle of categorical duality (see Section II.14). The principle only states that if Σ is a statement about a category **C**, then Σ^{op} is *universally* true if Σ is. For a *particular* category, it may very well happen that Σ is true but Σ^{op} is not.

15. Two Examples

Let $S = \{a, b, c\}$, $f: S \to \mathbb{R}$ with $f(a) = 0$, $f(b) = 1$, $f(c) = 2$; i.e., $S/R_f = \{\{a\}, \{b\}, \{c\}\}$, and $g: S \to \mathbb{R}$ with $g(a) = 0$, $g(b) = g(c) = 1$; i.e., $S/R_g = \{\{a\}, \{b, c\}\}$. Let $\phi:(S, \{f\}) \to (S, \{g\})$ be the identity on S, and $\phi f = g$. So diagrammatically we have

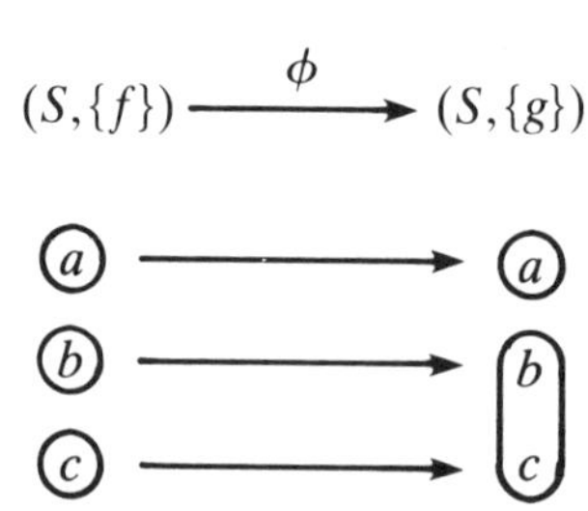

Clearly, ϕ is both mono and epi. Equally obviously, there is only one way to define ϕ^{-1}—namely, ϕ^{-1} is the identity on S, and $\phi^{-1} g = f$. But such ϕ^{-1} would not be acceptable as an **S**-morphism because b and c are related in $(S, \{g\})$ and not related in $(S, \{f\})$. So this example shows that, in **S**,

(i) mono and epi $\not\Rightarrow$ isomorphism;
(ii) mono $\not\Rightarrow$ split mono (hence mono $\not\Rightarrow$ equalizer);
(iii) epi $\not\Rightarrow$ split epi.

Further, in this example, since ϕ is mono, any $\psi_1, \psi_2 : (X, H) \to (S, \{f\})$ with $\phi \circ \psi_1 = \phi \circ \psi_2$ implies $\psi_1 = \psi_2$. But $\text{coeq}(\psi_1, \psi_1) = 1_{(S,\{f\})} \neq \phi$, so this example also shows that, in **S**,

(iv) epi $\Rightarrow$ coequalizer.

As a second example, consider $S = \{a, b, c, d\}$, $S/R_i = \{\{a\}, \{b\}, \{c\}, \{d\}\}$, $S/R_f = \{\{a, b\}, \{c\}, \{d\}\}$, $S/R_g = \{\{a\}, \{b, c\}, \{d\}\}$, $S/R_h = \{\{a, b, c\}, \{d\}\}$; and $\psi_1 : (S, \{i\}) \to (S, \{f, g\})$ with identity on S, $\psi_1 i = f$, $\psi_2 : (S, \{i\}) \to (S, \{f, g\})$ with identity on S, $\psi_2 i = g$, $\phi : (S, \{f, g\}) \to (S, \{h\})$ with identity on S, $\phi f = \phi g = h$. Diagrammatically,

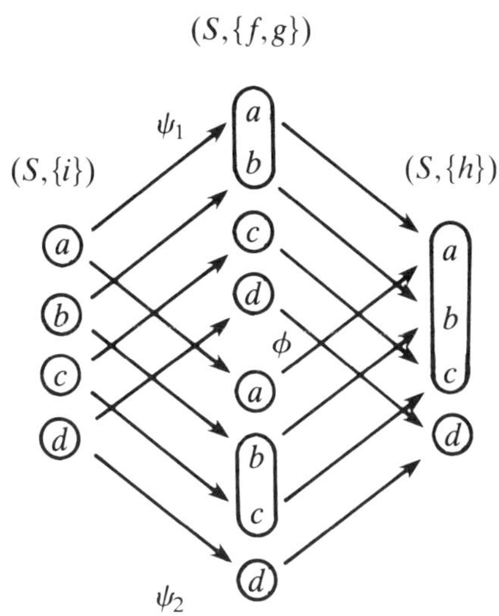

It is easy to check that $\phi = \text{coeq}(\psi_1, \psi_2)$. There can be no $\psi : (S, \{h\}) \to (S, \{f, g\})$ such that $\phi \circ \psi = 1_{(S,\{h\})}$, because we must have either $\psi h = f$ or $\psi h = g$, neither of which is acceptable for ψ to be an **S**-morphism. This shows that ϕ is not split epi. Thus in **S**,

(v) coequalizer $\Rightarrow$ split epi.

E. Images and Subobjects

16. Image Factorization

The categorical view of the *image* of a function f is as a factorization $f = i \circ p$, where p is surjective and i is injective. Two of the many possible such views in an arbitrary category are as follows. Given a morphism $f: A \to B$ in **C**, an *epi-equalizer factorization* of f is $f = i \circ p$ with p an epi and i an equalizer. The dual concept is a *coequalizer-mono factorization* $f = i \circ p$ with p a coequalizer and i a mono. It is a general theorem that epi-equalizer [and coequalizer-mono] factorizations are unique up to isomorphism. As a corollary, f is an isomorphism if and only if f is both an equalizer and epi and if and only if f is both a coequalizer and mono. (Note that a morphism that is epi and mono need not be an isomorphism, as we saw in Section III.15 for the case of **S**.)

We say **C** *has epi-equalizer* [dually, *has coequalizer-mono*] *factorizations* if every morphism in **C** has an epi-equalizer [respectively, a coequalizer-mono] factorization. The category **Ens** has epi-equalizer and coequalizer-mono factorizations and they both coincide with surjective–injective factorizations.

As a second example, the category **Top** of topological spaces and continuous functions has epi-equalizer and coequalizer-mono factorizations. Given a continuous function $f: A \to B$ with image factorization $f = i \circ p$ at the level of **Ens** we can provide $f(A)$ with the subspace topology induced by i, then (p, i) is an epi-equalizer factorization, or we can provide $f(A)$ with the quotient topology induced by p, in which case (p, i) is a coequalizer-mono factorization. The two image factorizations are clearly not in general homeomorphic.

The category **S**, also, has epi-equalizer and coequalizer-mono factorizations. The diagram

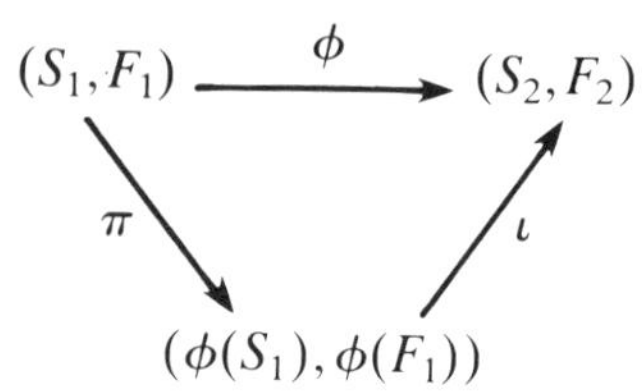

with $\pi = \phi$ on (S_1, F_1) and $\iota =$ inclusion is clearly an epi-equalizer factorization $\phi = \iota \circ \pi$ of ϕ (see Section III.10 that ι may not be well defined). Consider the equivalence relation $R = \{(s, s') \in S_1 \times S_1 : \phi(s) = \phi(s')\}$ on S_1 and $R = \{(f, f') \in F_1 \times F_1 : \phi f = \phi f'\}$ on F_1; the diagram

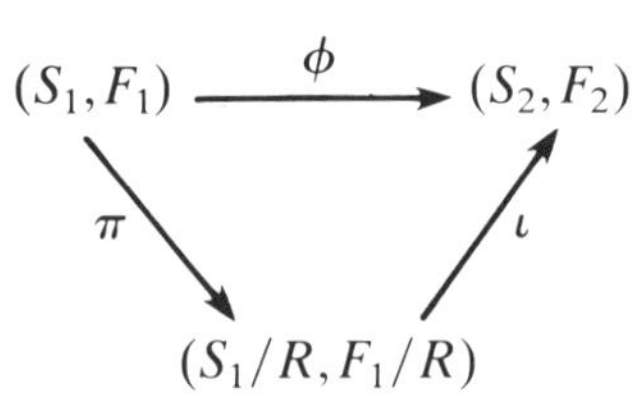

with $\pi =$ natural projection (where $F_1/R \subset \mathbb{R}^{S_1/R}$ is to be interpreted as in the construction of coequalizers in Section III.12) and ι defined by $\iota(s)_R = \phi(s)$,

$\iota(f)_R = \phi f$ (since sRs' iff $\phi s = \phi s'$, fRf' iff $\phi f = \phi f'$, ι is well defined) is then a coequalizer-mono factorization of ϕ.

Note that although both the epi-equalizer and coequalizer-mono factorizations are unique up to isomorphism, the two factorizations are *not* necessarily isomorphic, as we saw in the case of **Top**. The following example shows this for **S** (same ϕ as in the first example of Section III.15):

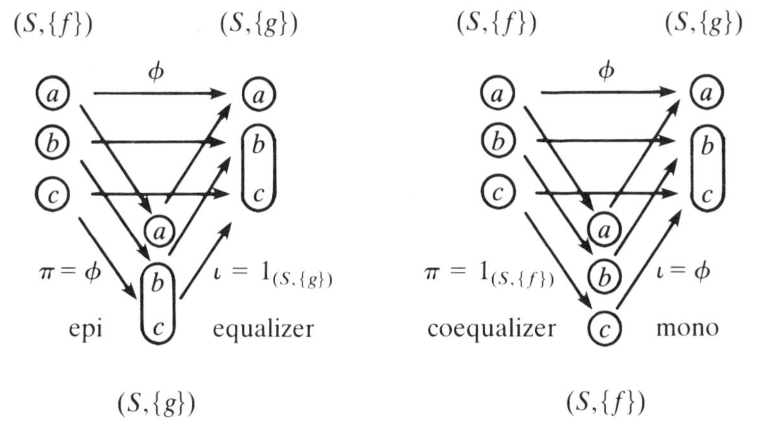

It is interesting to compare the image factorizations in **Top** and **S**.

17. Subobjects

In **Ens** (and in most of the categories studied in linear algebra), since a mono splits and all three definitions of monomorphism coincide with injection, there is only one way to define a subobject—namely, we say A is a subset of B (up to isomorphism, of course) if there is an injection $f: A \to B$. In a category **C**, the most common definition of a subobject of an object B is (the isomorphism class of) an object A with a monomorphism $f: A \to B$.

In **S**, as we saw, there are two distinct types of monomorphisms—namely, that of an equalizer (= split mono) and that of a mono (= injective morphism). We shall say that $\phi: (S_1, F_1) \to (S_2, F_2)$ is an **S**-*subsystem* (or simply (S_1, F_1) is a subsystem of (S_2, F_2)) if ϕ is an equalizer, and it is an *S-monosubobject* (or simply (S_1, F_1) is a monosubobject of (S_2, F_2)) if ϕ is mono. So a subsystem is a monosubobject but not vice versa. Note that *subsystem* implies that for each $f \in F_1$ $S_1/R_f \cong \phi(S_1)/R_{\phi f}$ ($\phi: S_1/R_f \to S_2/R_{\phi f}$ is one-to-one), i.e., $f(s) = f(s')$ if and only if $(\phi f)(\phi s) = (\phi f)(\phi s')$; whereas *monosubobject* does not have this "backward implication" ($\phi: S_1/R_f \to S_2/R_{\phi f}$ is not necessarily one to one). A subsystem, therefore, is the appropriate subobject of a system that preserves most of its structures.

18. A Partial Order in **S**

DEFINITION. Let (S_1, F_1) and (S_2, F_2) be **S**-objects. Then $(S_1, F_1) \leq (S_2, F_2)$ if

(i) (S_1, F_1) is a monosubobject of (S_2, F_2), i.e., there is a mono $\phi: (S_1, F_1) \to (S_2, F_2)$;
(ii) S_1 and F_1 are finite sets.

It is clear that $\leq$ is reflexive on *finite* **S**-*objects*, i.e., on (S, F), where S and F are finite sets, and that $\leq$ is transitive. Now suppose $(S_1, F_1) \leq (S_2, F_2)$ (with mono ϕ) and $(S_2, F_2) \leq (S_1, F_1)$ (with mono ψ), then $\phi: S_1 \to S_2$ and $\phi: F_1 \to F_2$ are onto because they are injections between finite sets of the same cardinality. So $\phi: S_1 \to S_2$ and $\phi: F_1 \to F_2$ are **Ens**-isomorphisms. Further, since ϕ and ψ are *order-preserving* on the linkage relations on the finite sets F_1 and F_2, we must have $S_1/R_f \cong S_2/R_{\phi f}$ for every $f \in F_1$. Thus (S_1, F_1) is isomorphic to (S_2, F_2) and so $\leq$ is antisymmetric (up to isomorphism). Therefore $\leq$ is a partial order on (the isomorphism class of) the finite **S**-objects.

What can we deduce from the statement $(S_1, F_1) \leq (S_2, F_2)$? First, we have $S_1 \rightarrowtail S_2$ and in fact for every $s \in S_1$ and every $f \in F_1$ $(s)_f \rightarrowtail (\phi s)_{\phi f}$. So there is a possibility of new states appearing in the whole set and/or in each equivalence class. This reflects *growth* in some respect. The possibility that $F_2 \sim \phi(F_1)$ is nonempty indicates the *emergence* of more observables as the system becomes "more advanced." In particular, there is the possibility that $sR_{F_1}s'$ in S_1 but there is a $g \in F_2 \sim \phi(F_1)$ such that $g(\phi s) \neq g(\phi s')$; so states indistinguishable before could be separated—a model of *differentiation*. On the other hand, it could happen that $f(s) \neq f(s')$ in (S_1, F_1) but $(\phi f)(\phi s) = (\phi f)(\phi s')$ in (S_2, F_2)—a model of *integration* or fusion. Also, since in this case 'distinct' states become indistinguishable, there is an indication of *decay*, or "loss of recognition abilities."

So it seems that with an appropriate totally ordered subset of this partially ordered set of finite systems, a model of the development–senescence process could be constructed. This will be a subject of study in Section VII.

As a final note, condition (ii) in the definition of $\leq$ (that S_1 and F_2 are to be *finite* sets) looks like a very severe mathematical restriction. But in mathematical modelling of natural systems, a finiteness restriction is not unrealistic: All we require is that the sets are finite, and there is no restriction on how small the sets have to be. So the sets could be singletons, have 10^{10} elements, or have 10^{100} elements and still be finite. After all, Sir James Jeans (1945) defined the universe as a gigantic machine whose future is inexorably fixed by its state at any given moment, that it is "a self-solving system of $6N$ simultaneous differential equations, where N is Eddington's number." Sir Arthur Eddington

(1939) asserted (perhaps with more poetry than truth) that $N = 2 \times 136 \times 2^{256}$ ($\sim 10^{79}$) is the total number of particles of matter in the universe. The point is that it is a *finite* number. Thus the set of states of a natural system is certainly finite at one time (this is not to be confused with the set of all *possible* states a system can have), and the set of observables on a system at one time is also clearly finite.

IV. DYNAMICAL SYSTEMS

> The world is a sojourn of all things. Time is a transversal of all ages. The essence of life is change.
>
> Lao Tse

A. Introduction

So far, the discussion has been entirely static. We now include dynamical considerations.

First, we shall restrict our attention to the dynamics of special physical systems called *meters*, through which our observables are defined. Then we shall see that general dynamics, i.e., a change of state in arbitrary systems through interaction of their states, is a corollary of the properties of meters and our fundamental hypothesis (Proposition 1, Section III.1) that all physical events can be represented by the evaluation of observables on states.

1. Meters and Observables

A *meter* M is a set in which a dynamics can be induced through interaction with the states of a system S. By hypothesis, M is in a particular reference state m_0, and M is measuring a specific observable f of S, in the following sense. On interaction of m_0 with a state $s \in S$, a change of state (i.e., a dynamics) is induced in the meter, carrying M (possibly asymptotically) to a state $m(s)$, which is labelled with a real number. The number assigned to $m(s)$ is then defined as the *value* $f(s)$ of the observable f of S. Two states $s, s' \in S$ produce the same response in M if and only if $f(s) = f(s')$. So the meter M can be regarded as interacting with the "reduced" states of S/R_f. Thus every meter defines an observable, and conversely for every observable we assume that there exists a meter in terms of which it could be defined.

2. Meter Dynamics

Since a change of state is a physical event, and since the very essence of the properties of a meter is that a change of state occurs as a result of interaction with other systems, it follows from Proposition 1 that every change of state can be regarded as being specified by the evaluation of observables on states and that every dynamical interaction between systems can be locally represented in terms of the dynamics induced on meters. Thus the reciprocal interaction of two meters sets up a situation similar to that of the "coordinate patches" in differential geometry and is the prototype for general dynamics.

A *dynamics* on a set of states S is usually defined as a one-parameter group (with time being the parameter) of automorphisms (i.e., bijections) on S, $\{T_t \in \mathscr{A}(S) : t \in \mathbb{R}\}$. These automorphisms are in general locally determined through identification of local dynamics with meter dynamics, measuring some observable f of another system S'. So there is a local correspondence $(S', \{f\}) \leftrightarrow (S, \{T_t\})$ between **S**-objects and "systems with dynamics."

In the next few sections we shall study some of the properties of the collection of "systems with dynamics," again this will be in the setting of category theory.

B. Dynamics and D-Objects

We shall first consider the objects of the category **D** of dynamical systems. It will turn out (Theorem IV.5) that our definition of dynamics is a generalization of the usual definition (as in Section IV.2 and Rosen, 1971).

3. *Dynamics*

DEFINITION. A *dynamics* on a set S is a mapping T from a subset of $S \times \mathbb{R}$ into S satisfying the following conditions:

(i) For every $x \in S$ there exist $a_x, b_x \in \bar{\mathbb{R}}$ (= the extended real numbers $[-\infty, +\infty]$) with $-\infty \leq a_x < 0 < b_x \leq +\infty$ such that $T(x,t)$ is defined if and only if $a_x < t < b_x$. (The notation $a_x[T]$ and $b_x[T]$ will be used when the dependence on T is emphasized.)

(ii) The "initial value property" $T(x,0) = x$ holds for all $x \in S$.

(iii) The "group property" $T(T(x,t_1),t_2) = T(x,t_1+t_2)$ holds if both $T(x,t_1)$ and also the left *or* right side of the equation is defined.

LEMMA. The preceding conditions (i), (ii), and (iii) are equivalent to (i), (iii), and

(ii*) T maps *onto* S.

Proof. Clearly, (ii) $\Rightarrow$ (ii*).

Conversely, assuming (i), (iii), and (ii*), take $x \in S$. Then (ii*) implies that there exists a $y \in S$ and there exists a $t \in (a_y, b_y) \subset \mathbb{R}$ such that $x = T(y,t)$. Now using (iii) and (i), $T(x,0) = T(T(y,t),0) = T(y,t+0) = T(y,t) = x$, proving (ii). $\square$

4. *Phase Space*

It follows from the lemma that the domain of T is nonempty if and only if S is nonempty. In fact, S is the projection onto the first coordinate of the domain of T. Also, $S =$ the range of T; we shall call S the *phase space* of T.

Each set S is the phase space of at least one dynamics, namely the *trivial dynamics* I_S defined by $I_S(x, t) = x$ for all $x \in S$ and all $t \in \mathbb{R}$.

As an example of a nontrivial dynamics, consider an autonomous differential equation (which, incidentally, started the study of dynamical systems)

$$dx/dt = f(x)$$

on an open subset S of $\mathbb{R}^k$, where $f: S \to \mathbb{R}^k$ is continuous and Lipschitz. Define a dynamics T on S as follows: for $x \in S$, let $y(t)$ be the unique solution of the equation that has $y(0) = x$ and maximal open interval (a_x, b_x) in $\mathbb{R}$ as domain; then set $T(x, t) = y(t)$ for $t \in (a_x, b_x)$. It is easy to check that T: dom $T \to S$ such defined satisfies the definition of a dynamics.

5. *Solutions and Translations*

Analogously to the preceding example, we can define *solutions* of a dynamics T on S as maps, for each $x \in S$, $y_x: (a_x, b_x) \to S$, $y_x(t) = T(x, t)$. Then it follows that $y_x(0) = x$, $y_{T(x,t_1)}(t_2) = y_x(t_1 + t_2)$ for all $x \in S$ and appropriate $t_1, t_2 \in \mathbb{R}$. The image of each y_x, i.e., the set $y_x(a_x, b_x) = \{y_x(t) : t \in (a_x, b_x)\}$ is called a *trajectory* of the dynamics. Note that $x_1 = y_{x_2}(t)$ if and only if $x_2 = y_{x_1}(-t)$, so for each $x \in S$ there is only one trajectory passing through it, namely, $y_x(a_x, b_x)$. This is the *unique trajectory property*, or the *principle of causality*. Clearly, the set of solutions $\{y_x : x \in S\}$ determines T uniquely.

On the other hand, fixing t and varying x, we can define *translations* of a dynamics T as, for each $t \in \mathbb{R}$, T_t: dom $T_t \to S$ where dom $T_t = \{x \in S : T(x, t)$ is defined$\}$, $T_t(x) = T(x, t)$. Then $T_0: S \to S$ with $T_0(x) = x$, $T_{t_2} \circ T_{t_1} = T_{t_1 + t_2}$, and $T_t(x) = y_x(t)$ (for all x and appropriate t). Again, it is clear that the translations $\{T_t : t \in \mathbb{R}\}$ determines T completely.

LEMMA. Each translation T_t: dom $T_t \to S$ is injective.

Proof. Suppose $T_t(x) = T_t(y)$, i.e., $T(x, t) = T(y, t)$. Then $x = T(x, 0) = T(T(x, t), -t) = T(T(y, t), -t) = T(y, 0) = y$. Note the second preceding equality uses the fact that both $T(x, t)$ and $T(x, 0)$ are defined, hence so is $T(T(x, t), -t)$; a similar comment goes to the fourth equality. [Here is an illustration of the importance of the *or* in condition (iii) of Definition IV.3. There is a profound difference if (iii) is altered to hold only when *all* terms concerned are defined.] □

THEOREM. If dom $T = S \times \mathbb{R}$, then each translation $T_t: S \to S$ is an automorphism (of sets, i.e., a bijection).

Proof. T_t is injective by the preceding lemma. If $y \in S$ then $x = T(y, -t) = T_{-t}(y)$ is defined, hence $y = T(y, 0) = T(T(y, -t), t) = T(x, t) = T_t(x)$. So T_t is surjective. $\square$

So when dom $T = S \times \mathbb{R}$, $\{T_t : t \in \mathbb{R}\}$ is a one-parameter group of (**Ens**-) automorphisms on S, i.e., a dynamics in the usual definition. So our definition of a dynamics can be considered as a generalization of the usual concept.

6. *Bounds*

DEFINITION. The a_x and b_x appearing in Definition IV.3(i) are called the *bounds of T at x.*

These bounds have the following interesting properties.

LEMMA. For every $x \in S$ and for every $t \in (a_x, b_x)$, $a_{T(x,t)} = a_x - t$ and $b_{T(x,t)} = b_x - t$.

Proof. For $x \in S$ and $t \in (a_x, b_x)$, $T(x, t)$ is defined. Now according to Definition IV.3(iii), $T(T(x,t), t')$ is defined if and only if $T(x, t + t')$ is defined [here again is where the *or* in (iii) is crucial]; then by (i), this means the statement

$$a_{T(x,t)} < t' < b_{T(x,t)}$$

is equivalent to

$$a_x < t + t' < b_x,$$

i.e.,

$$a_x - t < t' < b_x - t.$$

Thus $a_{T(x,t)} = a_x - t$ and $b_{T(x,t)} = b_x - t$. $\square$

COROLLARY 1. If $x \in S$ and either $a_x > -\infty$ or $b_x < +\infty$, then the trajectory $\gamma_x(a_x, b_x) = \{T(x,t): a_x < t < b_x\}$ has the cardinality of the continuum.

Proof. If $a_x > -\infty$ [respectively, $b_x < +\infty$], then all $a_{T(x,t)} = a_x - t$ [respectively, all $b_{T(x,t)} = b_x - t$] are distinct for t varying over (a_x, b_x). So all $T(x,t)$ as t varies over (a_x, b_x) must be distinct. $\square$

COROLLARY 2. If T is a dynamics on a countable set S, then dom $T = S \times \mathbb{R}$. $\square$

This last corollary is particularly interesting: on a countable (including finite) phase space S, the generalized and the usual definitions of *dynamics* coincide.

7. *Invariance and Relative Dynamics*

DEFINITION. If T is a dynamics on a set S, then $X \subset S$ is *invariant under* T (or simply T-*invariant*) if $T(X, \mathbb{R}) \subset X$ (i.e., for every $x \in X$ and for all $t \in (a_x, b_x)$, $T(x, t) \in X$).

Since $X = T(X, \{0\}) \subset T(X, \mathbb{R})$, the definition is equivalent to $T(X, \mathbb{R}) = X$. Note that S itself, the empty set, and any trajectory of T are all invariant under T.

It is clear that if A is a T-invariant subset of S, then the restriction of T to $(A \times \mathbb{R}) \cap \operatorname{dom} T$ is a dynamics T' on A, and for $x \in A$, $a_x[T'] = a_x[T]$, $b_x[T'] = b_x[T]$.

On the other hand, however, if A is a subset of S and if the restriction T' of a dynamics T on S to some subset of $(A \times \mathbb{R}) \cap \operatorname{dom} T$ is a dynamics on A (we shall call T' a *relative dynamics* on A induced by T on S), A is not necessarily T-invariant. All is required here is that for $x \in A$,

$$a_x[T] \leq a_x[T'] < 0 < b_x[T'] \leq b_x[T]$$

and that T' satisfies Definition IV.3. (In particular, $a_x[T']$ and $b_x[T']$ satisfy Lemma IV.6.) Here $\operatorname{dom} T' \subset (A \times \mathbb{R}) \cap \operatorname{dom} T$ but not necessarily equal.

So we have

$$A \text{ is } T\text{-invariant} \Rightarrow T \text{ induces a relative dynamics on } A,$$

but not conversely.

8. *Dynamical System*

DEFINITION. A **D**-*object*—a *dynamical system*—is a pair (S, D), where S is a set (of states, the phase space) and D is a set of dynamics on S.

We always assume that the trivial dynamics I_S is in D, so in particular D is always nonempty; but again for brevity we shall sometimes omit listing I_S in specific examples. Note that if a dynamics is considered to be imposed on S through interactions with the states of other systems, then I_S can be considered as being imposed on S through interaction with the (only) state in S/R_0 (where 0 is again the zero observable as in Section III).

C. D-Morphisms

We have studied the objects of **D** in some detail. The next question is, "What are the morphisms?" How shall we define a **D**-morphism so that **D**-isomorphic objects are identical in their dynamical behaviour?

9. **D**-*Morphism*

DEFINITION. $\phi \in \mathbf{D}((S_1, D_1), (S_2, D_2))$ if it is a pair of functions $\phi: S_1 \to S_2$ and $\phi: D_1 \to D_2$ (with $\phi I_{S_1} = I_{S_2}$), and such that for every $T \in D_1$ the diagram

$$\begin{array}{ccc} \operatorname{dom} T & \xrightarrow{\phi \times 1_{\mathbb{R}}} & \operatorname{dom} \phi T \\ {\scriptstyle T}\downarrow & & \downarrow{\scriptstyle \phi T} \\ S_1 & \xrightarrow[\phi]{} & S_2 \end{array}$$

commutes [i.e., for every $x \in S_1$ and for all $t \in (a_x[T], b_x[T])$, $\phi(T(x,t)) = \phi T(\phi x, t)$]. Note this implies that for all $T \in D_1$ and for all $x \in S_1$ $a_{\phi x}[\phi T] \le a_x[T] < 0 < b_x[T] \le b_{\phi x}[\phi T]$.

Also note that the diagram

$$\begin{array}{ccc} S_1 \times \mathbb{R} & \xrightarrow{\phi \times 1_{\mathbb{R}}} & S_2 \times \mathbb{R} \\ {\scriptstyle I_{S_1}}\downarrow & & \downarrow{\scriptstyle \phi I_{S_1} = I_{S_2}} \\ S_1 & \xrightarrow[\phi]{} & S_2 \end{array}$$

trivially commutes; so the inclusion of I_S into the dynamics does not create any problems.

It follows from the definition of a **D**-morphism that ϕT induces a relative dynamics on $\phi(S_1)$ with relative bounds $a_x[T]$ and $b_x[T]$ at ϕx—in fact, if $T \in D_1$ induces a relative dynamics on $A \subset S_1$, then ϕT induces a relative dynamics on $\phi(A) \subset S_2$. In particular, this last statement holds for $A \subset S_1$ invariant under T (see Section IV.7). Note, however, that A invariant under T does not imply ϕA invariant under ϕT.

Trivially, $1_{(S,D)}: (S, D) \to (S, D)$ sending each $x \in S$ and each $T \in D$ to itself is the identity morphism in $\mathbf{D}((S, D), (S, D))$.

10. *Composition*

Composition of **D**-morphisms is defined via compositions of functions on the states and on the dynamics. The commutativity condition is represented by the self-explanatory diagram

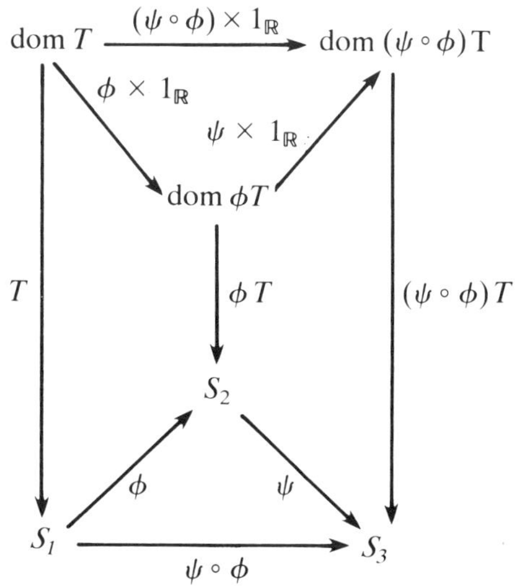

It is now easy to check that with the preceding definitions, **D** is a category.

11. Isomorphisms

If $\phi:(S_1, D_1) \to (S_2, D_2)$ and $\psi:(S_2, D_2) \to (S_1, D_1)$ are such that $\psi \circ \phi = 1_{(S_1,D_1)}$ and $\phi \circ \psi = 1_{(S_2,D_2)}$, then it follows that $\phi: S_1 \to S_2$ and $\phi: D_1 \to D_2$ are bijections. Further, with $T \in D_1$ and $\phi T = U \in D_2$ (so $\psi U = T$), we have

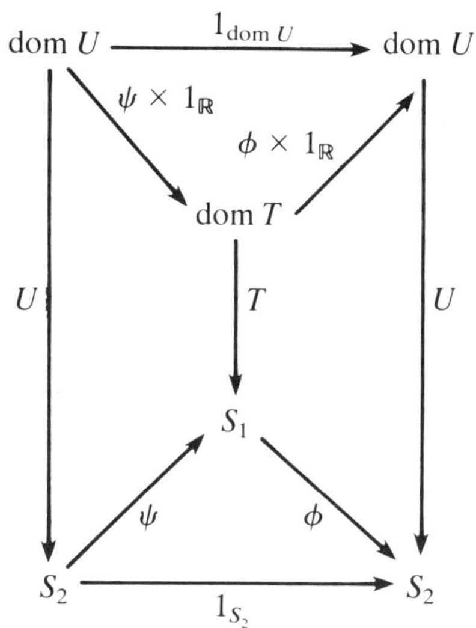

In particular we see that $U = \phi T = \phi \circ T \circ (\psi \times 1_{\mathbb{R}})$, and for each $x \in S_1$ $a_{\phi x}[\phi T] = a_x[T]$ and $b_{\phi x}[\phi T] = b_x[T]$, hence dom $\phi T \cong$ dom T.

So we see that to say (S_1, D_1) is isomorphic to (S_2, D_2) means that they are the "same" systems down to the fact that corresponding dynamics have isomorphic domains. When dynamical systems are considered as meters, two isomorphic systems are abstractly identical in the sense that they are measuring the same observables. There is a "dictionary" $\phi \leftrightarrow \psi$ that converts one system to the other. This reparametrization can be physically realized as a scale conversion.

D. Constructions in D

12. Products

Let $\{(S_i, D_i): i \in I\}$ be a family in **D**. Then, letting the phase space be $S = \Pi S_i$, for each choice of $(T_i: i \in I) \in \Pi D_i$, it is natural to define a dynamics T on S—the *product dynamics* of $\{T_i: i \in I\}$—by

$$T((x_i)_{i \in I}, t) = (T_i(x_i, t))_{i \in I}. \tag{1}$$

One can easily check that the canonical projection maps π_j with $x = (x_i)_{i \in I} \mapsto x_j$, $T = (T_i)_{i \in I} \mapsto T_j$ are indeed **D**-morphisms and this construction satisfies the universal property for categorical products.

There is, however, one problem: In Eq. (1), it is required that $T_i(x_i, t)$ be defined for all $i \in I$. It is possible, when I is infinite, that for some choice of the x_i's, we have $a_{(x_i)}[T] = \sup\{a_{x_i}[T_i]: i \in I\} = 0$ and/or $b_{(x_i)}[T] = \inf\{b_{x_i}[T_i]: i \in I\} = 0$. Then in this case T cannot be a dynamics (we need $a < 0 < b$).

So when is $T = (T_i: i \in I)$ a dynamics on S? For $a_{(x_i)}[T] = \sup\{a_{x_i}[T_i]: i \in I\}$ to be negative, it is sufficient that $a_{x_i}[T_i] = -\infty$ for all but finitely many $i \in I$; similarly, for $b_{(x_i)}[T] = \inf\{b_{x_i}[T_i]: i \in I\}$ to be positive, it is sufficient that $b_{x_i}[T_i] = +\infty$ for all but finitely many $i \in I$. Thus when dom $T_i = S_i \times \mathbb{R}$ for all but finitely many i, T is a dynamics on S. In fact, it turns out that this is also a necessary condition:

THEOREM. A necessary and sufficient condition for T defined by Eq. (1) to be a dynamics on $S = \Pi S_i$ is that dom $T_i = S_i \times \mathbb{R}$ for all but finitely many $i \in I$.

Proof. The sufficiency is proved above. As for the necessity, suppose $J \subset I$ is a countably infinite subset such that dom $T_j \neq S_j \times \mathbb{R}$ for $j \in J$. Then for each $j \in J$ there exists $y_j \in S_j$ with finite bound(s). Assume, without loss of generality, that $b_{y_j}[T_j] < +\infty$, and that $J = \mathbb{N}$ (i.e., since J is countably infinite, let the corresponding element in $\mathbb{N}$ in the enumeration of J to

$j \in J$ be j). Define $t_j = \max\{0, b_{y_j}[T_j] - 1/j\}$ and $x_j = T_j(y_j, t_j)$. Then by Lemma IV.6

$$b_{x_j}[T_j] = b_{T_j(y_j,t_j)}[T_j] = b_{y_j}[T_j] - t_j \leq 1/j.$$

Now pick $x_i \in S_i$ arbitrarily for $i \notin J$, and let $x = (x_i : i \in I)$. Then

$$b_x[T] = \inf_{i \in I} b_{x_i}[T_i] \leq \inf_{j \in J} b_{x_j}[T_j] \leq \inf_{j \in J} 1/j = 0.$$

Hence T cannot be a dynamics on S. $\square$

So the **D**-product of the family $\{(S_i, D_i) : i \in I\}$ is defined as (S, D), where $S = \Pi S_i$, and $D =$ the subset of ΠD_i such that $T = (T_i : i \in I) \in D$ if and only if dom $T_i = S_i \times \mathbb{R}$ for all but finitely many $i \in I$. Note that $I_S = (I_{S_i} : i \in I)$ (the trivial dynamics on S) is such that dom $I_{S_i} = S_i \times \mathbb{R}$ for all $i \in I$, hence $I_S \in D$ always. Also, note that $D = \Pi D_i$ when I is finite.

It is interesting to note the resemblance of the definition of *product dynamics* to that of the *product topology*. Recall that in a product topological space $X = \Pi X_i$, sets of the form $A = \Pi A_i$, where each A_i is open in X_i and $A_i = X_i$ for all but finitely many i, form a base for the product topology.

Since **D** has products, it has a final object. The final object in **D** is $(1, \{I_1\})$, where 1 is a singleton set. For any dynamical system (S, D), the unique **D**-morphism $\phi : (S, D) \to (1, \{I_1\})$ is clearly the one sending all $x \in S$ to 1 and all $T \in D$ to I_1.

13. *Coproducts*

Coproducts of dynamical systems can also be defined. Let $\{(S_i, D_i) : i \in I\}$ be a family in **D**, let the coproduct phase space be $S = \amalg S_i = \cup(\{i\} \times S_i)$, and let the coproduct dynamics D be $\amalg D_i$. A dynamics $(i, T) \in D$ is defined by

$$(i, T)((j, x), t) = \begin{cases} (i, T(x, t)) & \text{if } j = i, \quad \text{for } a_x[T] < t < b_x[T] \\ (j, x) & \text{if } j \neq i, \quad \text{for } t \in \mathbb{R}. \end{cases}$$

So, roughly, (i, T) restricted to $\{i\} \times S_i$ is T on S_i and (i, T) restricted to $\amalg S_j : j \neq i$ is the trivial dynamics. Note we have $a_{(i,x)}[(i, T)] = a_x[T]$, $b_{(i,x)}[(i, T)] = b_x[T]$, and for $j \neq i$ $a_{(j,x)}[(i, T)] = -\infty$, $b_{(j,x)}[(i, T)] = +\infty$, i.e., $\text{dom}(i, T) = (\{i\} \times \text{dom } T) \bigcup \amalg (S_j \times \mathbb{R} : j \neq i)$.

The canonical injection maps $\iota_j : (S_j, D_j) \to (S, D)$ with $x \mapsto (j, x)$ and $T \mapsto (j, T)$ are such that

$$\iota_j(T(x, t)) = (j, T(x, t)) = (j, T)((j, x), t) = \iota_j T(\iota_j(x), t),$$

i.e., they are **D**-morphisms. One easily checks that this construction satisfies the universal property for categorical coproducts.

The initial object of **D** is $(\varnothing, \{I_\varnothing\})$. For any dynamical system (S, D), the unique **D**-morphism from $(\varnothing, \{I_\varnothing\})$ to (S, D) is the inclusion of $\varnothing$ in S with $I_\varnothing \mapsto I_S$.

E. Hierarchy of D-Morphisms and Image Factorizations

14. Monomorphisms

The hierarchy for **D**-monomorphisms is

$$\text{split mono} \Rightarrow \text{equalizer} \Leftrightarrow \text{mono} \Leftrightarrow \text{injection on } S \text{ and } D.$$

It is clear that a **D**-morphism that is injective on both S and D is mono. To show the converse, let $\phi:(S_1, D_1) \to (S_2, D_2)$ be a mono. Suppose there are distinct x and y in S_1 with $\phi x = \phi y$. Then ψ_1, $\psi_2:(S_1, D_1) \to (S_1, D_1)$, with ψ_1 sending all of S_1 to x, ψ_2 sending all of S_1 to y, and $\psi_1, \psi_2 =$ identity on D_1, are **D**-morphisms with $\phi \circ \psi_1 = \phi \circ \psi_2$ but $\psi_1 \neq \psi_2$, a contradiction. So ϕ must be injective on S_1. Likewise suppose there are distinct T and T' in D_1 and $\phi T = \phi T'$. There are two cases to consider: either $T(x, t) = T'(x, t)$ whenever $(x, t) \in \operatorname{dom} T \bigcap \operatorname{dom} T'$ (hence $\operatorname{dom} T \neq \operatorname{dom} T'$), or there exists $(x_0, t_0) \in \operatorname{dom} T \bigcap \operatorname{dom} T'$ for which $T(x_0, t_0) \neq T'(x_0, t_0)$. In the former case defined U to be the restriction of T (and T') to $\operatorname{dom} T \bigcap \operatorname{dom} T'$, then ψ_1, $\psi_2:(S_1, \{U\}) \to (S_1, D_1)$, with ψ_1, $\psi_2 =$ identity on S_1, $\psi_1 U = T$, $\psi_2 U = T'$, are **D**-morphisms with $\phi \circ \psi_1 = \phi \circ \psi_2$ but $\psi_1 \neq \psi_2$. In the latter case, let $S = \operatorname{dom} T_{t_0} \bigcap \operatorname{dom} T'_{t_0}$; then ψ_1, $\psi_2:(S, \{I_S\}) \to (S_1, D_1)$, with $\psi_1(x) = T(x, t_0)$, $\psi_2(x) = T'(x, t_0)$, $\psi_1 I_S = \psi_2 I_S = I_{S_1}$, are **D**-morphisms such that $\phi \circ \psi_1 = \phi \circ \psi_2$, because $(\phi \circ \psi_1)(x) = \phi(T(x, t_0)) = \phi T(\phi x, t_0) = \phi T'(\phi x, t_0) = \phi(T'(x, t_0)) = (\phi \circ \psi_2)(x)$. Note that $x_0 \in S$ is such that $\psi_1(x_0) = T(x_0, t_0) \neq T'(x_0, t_0) = \psi_2(x_0)$, hence $\psi_1 \neq \psi_2$. In both cases we are led to contradictions. So ϕ must be injective on D_1 as well. Thus a **D**-mono is injective on both the sets of states and dynamics.

To show that a mono $\phi:(S_1, D_1) \to (S_2, D_2)$ in **D** is an equalizer $\phi = \operatorname{eq}(\psi_1, \psi_2)$, ψ_1 and ψ_2 are defined as follows. First, let $S = \{0\} \times \phi(S_1) \bigcup \{1,2\} \times (S_2 \sim \phi(S_1))$—i.e., S is the disjoint union of one copy of $\phi(S_1)$ and two copies of $S_2 \sim \phi(S_1)$—and similarly let $D = \{0\} \times \phi(D_1) \bigcup \{1,2\} \times (D_2 \sim \phi(D_1))$. The members of D are defined in Table I. One can easily check that D is indeed a collection of dynamics on S [e.g., each $(0, T)$ is a dynamics on S because T induces a relative dynamics on $\phi(S_1)$ with the given relative bounds]. Now let ψ_1, $\psi_2:(S_2, D_2) \to (S, D)$ be

$$\psi_i(x) = \begin{cases} (0, x), & x \in \phi(S_1) \\ (i, x), & x \notin \phi(S_1) \end{cases} \qquad \psi_i T = \begin{cases} (0, T), & T \in \phi(D_1) \\ (i, T), & T \notin \phi(D_1) \end{cases}$$

Table I

Definition of Dynamics in D[a]

	$((0, x), t)$	$((1, x), t)$	$((2, x), t)$
$(0, T)$	$(0, T(x,t))$[b]	$(1, T(x,t))$[c]	$(2, T(x,t))$[c]
$(1, T)$	$(0, T(x,t))$, if $T(x,t) \in \phi(S_1)$; $(1, T(x,t))$, if $T(x,t) \notin \phi(S_1)$.[c]	$(0, T(x,t))$, if $T(x,t) \in \phi(S_1)$; $(1, T(x,t))$, if $T(x,t) \notin \phi(S_1)$.[c]	$(2, x)$[d]
$(2, T)$	$(0, T(x,t))$, if $T(x,t) \in \phi(S_1)$; $(2, T(x,t))$, if $T(x,t) \notin \phi(S_1)$.[c]	$(1, x)$[d]	$(0, T(x,t))$, if $T(x,t) \in \phi(S_1)$; $(2, T(x,t))$, if $T(x,t) \notin \phi(S_1)$.[c]

[a] The value of the dynamics i at the point j is given in the (row i, column j)-position. $T \in D_2$ and $x \in S_2$.
[b] For $a_y[U] < t < b_y[U]$, where $\phi y = x$, $\phi U = T$.
[c] For $a_x[T] < t < b_x[T]$.
[d] For $-\infty < t < +\infty$.

for $i = 1, 2$. One can check that ψ_1, ψ_2 are indeed **D**-morphisms, that $\phi(S_1) \subset S_2$ and $\phi(D_1) \subset D_2$ are the sets on which ψ_1 and ψ_2 agree, and $\psi_1 \circ \phi = \psi_2 \circ \phi$. Since $S_1 \cong \phi(S_1)$ and $D_1 \cong \phi(D_1)$, $\phi: (S_1, D_1) \to (S_2, D_2)$ is (up to isomorphism) the equalizer of ψ_1 and ψ_2. (It is clear, as in **S**, that the equalizer of a pair of morphisms in **D** is the inclusion map of the subsystem on which the morphisms agree.)

Finally, we show that an equalizer is not necessarily split mono. Let $S_1 = \{a, b\}$ and $S_2 = \{a, b, c\}$. Let $T: S_1 \times \mathbb{R} \to S_1$ be defined by $T(a, 0) = T(b, t) = a$, $T(b, 0) = T(a, t) = b$ for $t \neq 0$. Let $U: S_2 \times \mathbb{R} \to S_2$ be defined by $U(a, 0) = U(b, t) = a$, $U(b,0) = U(a, t) = b$ for $t \neq 0$, and $U(c, t) = c$ for all t. It is easy to see that T and U are dynamics on S_1 and S_2, respectively. Now let $\phi: (S_1, \{T\}) \to (S_2, \{U\})$ be $\phi(a) = a$, $\phi(b) = b$, and $\phi T = U$. Then ϕ is a **D**-morphism.

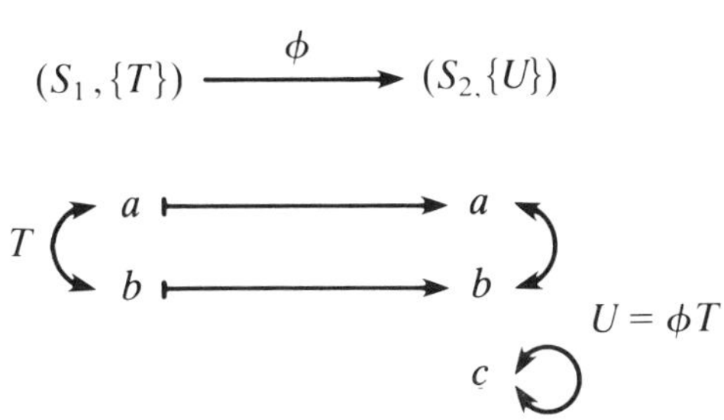

Any definition of ϕ^{-1} must have $\phi^{-1}(a) = a$, $\phi^{-1}(b) = b$, and $\phi^{-1}U = T$; but either $\phi^{-1}(c) = a$ or $\phi^{-1}(c) = b$ would be inconsistent: if $\phi^{-1}(c) = a$ then for $t \neq 0$, $\phi^{-1}(U(c,t)) = \phi^{-1}(c) = a$ but $T(\phi^{-1}(c), t) = T(a,t) = b$, hence the diagram

$$\begin{array}{ccc} S_2 \times \mathbb{R} & \xrightarrow{\phi^{-1} \times 1_{\mathbb{R}}} & S_1 \times \mathbb{R} \\ {\scriptstyle U}\downarrow & & \downarrow{\scriptstyle T} \\ S_2 & \xrightarrow[\phi^{-1}]{} & S_1 \end{array} \qquad \text{at} \qquad \begin{array}{ccc} (c,t) & \longmapsto & (a,t) \\ \downarrow & (t \neq 0) & \downarrow \\ c & \longmapsto & a \neq b \end{array}$$

is not commutative; similarly, for $\phi^{-1}(c) = b$. Thus ϕ does not split.

So we have established, in **D**,

$$\text{split mono} \Rightarrow \text{equalizer} \Leftrightarrow \text{mono} \Leftrightarrow \text{injection on } S \text{ and } D.$$

We can now define a **D**-*subobject*—a *dynamical subsystem*—(S_1, D_1) of (S_2, D_2) to be one such that there exists a mono (= equalizer = injection) ϕ from (S_1, D_1) to (S_2, D_2). Note that in this case, since $\phi(S_1) \cong S_1$, each $T \in D_1$ can be considered as the relative dynamics on S_1 induced by the dynamics ϕT on S_2.

15. Image Factorization Systems

Before we look at the hierarchy of **D**-epimorphisms, it would be useful to consider some more general concepts from category theory.

DEFINITION. An *image factorization system* for a category **C** is a pair $(\mathscr{E}, \mathscr{M})$, where $\mathscr{E}$ and $\mathscr{M}$ are classes of **C**-morphisms such that

(i) $\mathscr{E}$ and $\mathscr{M}$ are subcategories of **C** [i.e., compositions of morphisms in $\mathscr{E}$ (respectively, $\mathscr{M}$) stay in $\mathscr{E}$ (respectively, $\mathscr{M}$)];

(ii) every element of $\mathscr{E}$ is an epi (possibly more; e.g., it could be a coequalizer) and every element of $\mathscr{M}$ is a mono (possibly more);

(iii) every isomorphism is in both $\mathscr{E}$ and $\mathscr{M}$;

(iv) every morphism f in **C** has a unique (up to isomorphism) $\mathscr{E}$–$\mathscr{M}$ factorization; i.e., there exist $e \in \mathscr{E}$ and $m \in \mathscr{M}$ such that $f = m \circ e$, and whenever $e' \in \mathscr{E}$ and $m' \in \mathscr{M}$ satisfy $f = m' \circ e'$, there exists a unique isomorphism ϕ with $e' = \phi \circ e$ and $m' \circ \phi = m$.

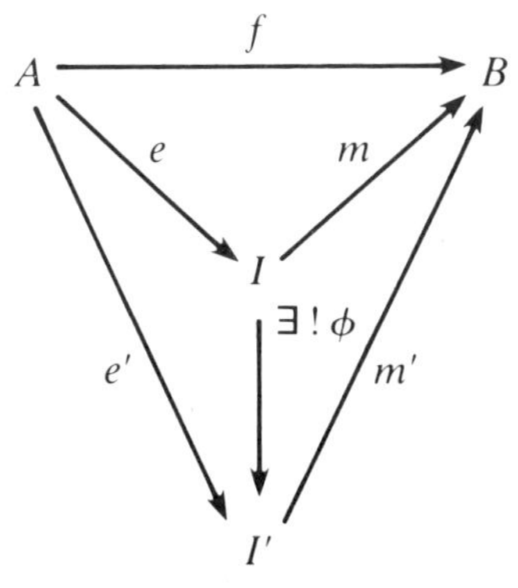

If **C** has epi-equalizer factorizations, then $(\mathscr{E}, \mathscr{M})$ is an image factorization system if $\mathscr{E}$ = all epis and $\mathscr{M}$ = all equalizers. Dually, if **C** has coequalizer-mono factorizations, then $\mathscr{E}$ = all coequalizers and $\mathscr{M}$ = all monos form an image factorization system for **C**.

A consequence of (iv) in the definition is that if a morphism is in both $\mathscr{E}$ and $\mathscr{M}$ then it is an isomorphism; i.e., the converse of (iii) holds.

THE DIAGONAL FILL-IN LEMMA. Given the commutative square

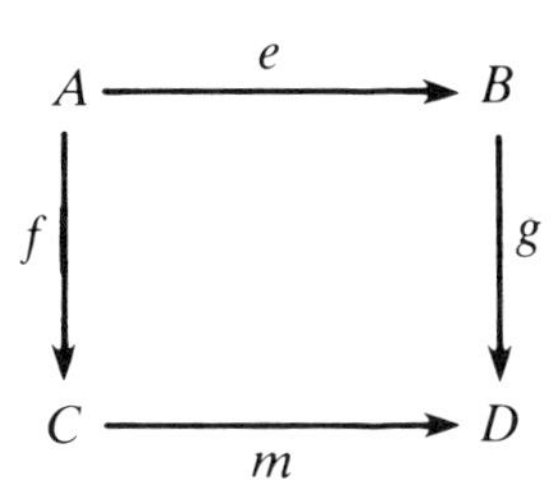

with $e \in \mathscr{E}$ and $m \in \mathscr{M}$, there exists a unique $h: B \to C$ with $f = h \circ e$ and $g = m \circ h$.

Proof. Consider the diagram

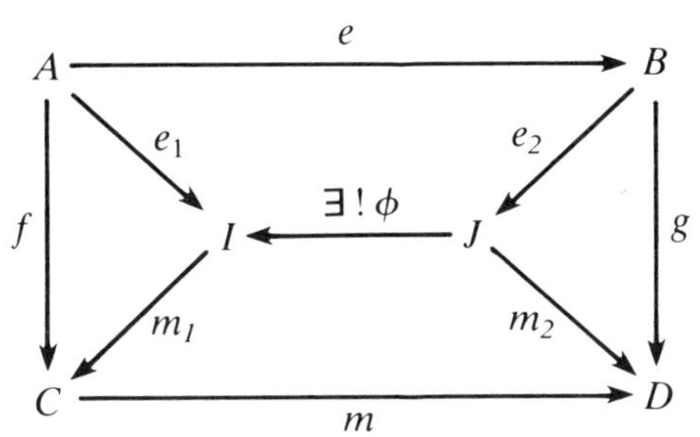

Define $h = m_1 \circ \phi \circ e_2$. □

16. Epimorphisms

The hierarchy for **D**-epimorphisms is

$$\text{split epi} \Rightarrow \text{coequalizer} \Leftrightarrow \text{epi} \Leftrightarrow \text{surjection on } S \text{ and } D.$$

It is clear that a **D**-morphism that is surjective onto S and D is epi. For the converse, let $\phi:(S_1, D_1) \to (S_2, D_2)$ be an epi. Suppose there exists $U \in D_2 \sim \phi(D_1)$. Let $S = S_2 \bigcup \{a, b\}$ where $a, b \notin S_2$, and $D = \{0, 1\} \times D_2$. $(0, T) \in D$ is defined on S as T on S_2 and identity on $\{a, b\}$. $(1, T) \in D$ is defined on S as T on S_2 and it "interchanges" a and b (i.e., $(1, T)(a, 0) = (1, T)(b, t) = a$ and $(1, T)(b, 0) = (1, T)(a, t) = b$ for $t \neq 0$). Then (S, D) is a **D**-object. Let $\psi_1:(S_2, D_2) \to (S, D)$ be defined by sending each $x \in S_2$ to itself and sending each $T \in D_2$ to $(0, T)$; and let $\psi_2:(S_2, D_2) \to (S, D)$ be defined by sending each $x \in S_2$ to itself and sending each $T \in D_2$ to $(0, T)$ *except* $\psi_2 \mathrm{U} = (1, \mathrm{U})$. Then one sees that ψ_1 and ψ_2 are **D**-morphisms with $\psi_1 \circ \phi = \psi_2 \circ \phi$ but $\psi_1 \neq \psi_2$, a contradiction. So ϕ must be onto D. If $S_2 \sim \phi(S_1)$ is nonempty, consider $\psi_1, \psi_2:(S_2, D_2) \to (\{0, 1\}, \{I_{\{0,1\}}\})$ with $\psi_1 = \chi_{\phi(S_1)}$ on S_2, $\psi_2 = \chi_{S_2}$ on S_2 and $\psi_1 T = \psi_2 T = I_{\{0,1\}}$ for all $T \in D_2$. Then since T induces a relative dynamics on $\phi(S_1)$ (because $T \in D_2 = \phi(D_1)$), ψ_1 is a **D**-morphism; ψ_2 is trivially a **D**-morphism. And $\psi_1 \circ \phi = \psi_2 \circ \phi$ with $\psi_1 \neq \psi_2$, a contradiction. So ϕ must also be onto S_2. Thus an epi is onto both S and D.

To show that an epi in **D** is a coequalizer, we shall use the diagonal fill-in lemma from the previous section. Let $\phi:(S_1, D_1) \to (S_2, D_2)$ be a **D**-morphism, let $\phi = m_1 \circ e_1$ be its epi-equalizer factorization, and let $\phi = m_2 \circ e_2$ be its coequalizer-mono factorization. Let $\mathscr{E}$ = all **D**-epis and $\mathscr{M}$ = all **D**-equalizers, hence $(\mathscr{E}, \mathscr{M})$ is an image factorization system for **D**. We have the commutative diagram

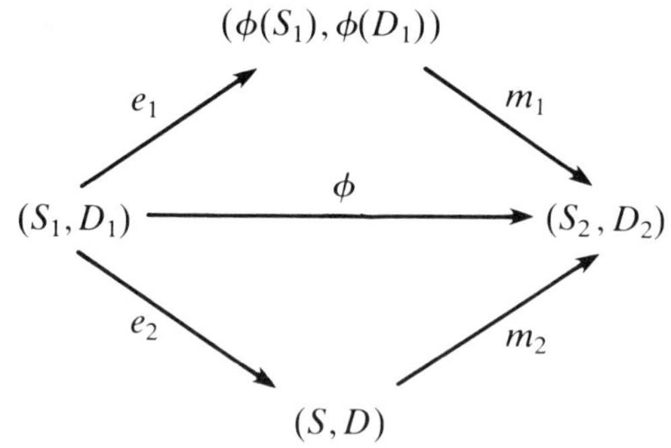

[It is clear that the "image object" in the epi-equalizer factorization is $(\phi(S_1), \phi(D_1))$.]

Now consider the pair (e_1, m_2). $e_1 \in \mathscr{E}$, and m_2 is a **D**-mono hence a **D**-equalizer (see Section IV.14), so $m_2 \in \mathscr{M}$. Thus by the diagonal fill-in lemma there exists a unique $g:(\phi(S_1), \phi(D_1)) \to (S, D)$ with $e_2 = g \circ e_1$ and $m_1 = m_2 \circ g$. On the other hand, the pair (e_2, m_1) is such that e_2 is a coequalizer and hence an epi (general hierarchy), so $e_2 \in \mathscr{E}$, and $m_1 \in \mathscr{M}$. Thus there exists a unique $h:(S, D) \to (\phi(S_1), \phi(D_1))$ with $e_1 = h \circ e_2$ and $m_2 = m_1 \circ h$. Now $e_2 = g \circ e_1 = g \circ h \circ e_2$ and $e_1 = h \circ e_2 = h \circ g \circ e_1$, so $g \circ h = 1_{(S,D)}$ and $h \circ g = 1_{(\phi(S_1), \phi(D_1))}$ (because e_1 and e_2 are epi), whence (S, D) is isomorphic to $(\phi(S_1), \phi(D_1))$ in **D**.

This shows that the epi-equalizer and coequalizer-mono factorizations in **D** are the same and so in particular an epi is a coequalizer.

The following simple example shows that a **D**-coequalizer is not necessarily a split epi. Let $S_1 = \{a, b\}$ and $S_2 = \{c\}$. Let T be the dynamics on S_1 defined by $T(a, 0) = T(b, t) = a$, $T(b, 0) = T(a, t) = b$ for $t \neq 0$. Then $\phi:(S_1, \{T\}) \to (S_2, \{I_{S_2}\})$ with $\phi(a) = \phi(b) = c$ and $\phi T = I_{S_2}$ is clearly epi, hence a coequalizer.

$$(S_1, \{T\}) \xrightarrow{\ \phi\ } (S_2, \{I_{S_2}\})$$

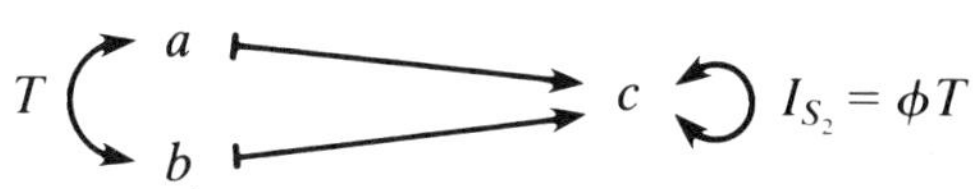

If ϕ^{-1} exists, it has to have $\phi^{-1}(c) = a$ or b. But one easily checks that neither case would yield a **D**-morphism.

So we have, in **D**,

$$\text{split epi} \Rightarrow \text{coequalizer} \Leftrightarrow \text{epi} \Leftrightarrow \text{surjection on } S \text{ and } D.$$

And, quite different from **S**, there is only one image factorization in **D**—epi-equalizer = coequalizer-mono = surjection-injection.

F. Dynamics and Equivalence Relations

17. Compatibility

DEFINITION. Let T be a dynamics on S. T is *compatible* with an equivalence relation R on S if xRy implies

(i) $T(x, t)$ is defined if and only if $T(y, t)$ is defined, and
(ii) $T(x, t) R T(y, t)$.

When dom $T = S \times \mathbb{R}$, condition (i) is of course superfluous and the compatibility condition reduces to the usual one for the usual definition of dynamics (cf. Rosen (1978)).

18. Quotient Dynamics

If T is compatible with the equivalence relation R on S, then T induces a dynamics T' on the set of 'reduced states' $S' = S/R$. For any $(x)_R \in S'$ and $t \in \mathbb{R}$, let $T'((x)_R, t) = (T(x,t))_R \in S'$ whenever $T(x,t)$ is defined, and leave $T'((x)_R, t)$ undefined otherwise.

To show that T' is a well-defined function, we have to show that it is independent of the equivalence class representative x. But for xRy, $T(x,t)$ is defined if and only if $T(y,t)$ is defined, and $T(x,t)RT(y,t)$, whence $(T(x,t))_R = (T(y,t))_R$. So T' is indeed independent of the choice of equivalence class representatives. Further, condition IV.17(i) actually implies that when xRy, $a_x[T] = a_y[T]$ and $b_x[T] = b_y[T]$, so we obtain the well-defined bounds for T' at $(x)_R$ as $a_{(x)_R}[T'] = a_x[T]$ and $b_{(x)_R}[T'] = b_x[T]$.

The initial value property $T'((x)_R, 0) = (T(x,0))_R = (x)_R$ clearly holds for T'. To show that T' is indeed a dynamics, it remains to verify the group property of IV.3(iii). Suppose $T'((x)_R, t_1) = (T(x,t_1))_R$ is defined (thus $T(x,t_1)$ is defined, in particular); then if $T'(T'((x)_R, t_1), t_2) = T'((T(x,t_1))_R, t_2)$ is defined, it is equal to $(T(T(x,t_1),t_2))_R = (T(x, t_1 + t_2))_R = T'((x)_R, t_1 + t_2)$, using the group property for T. Similarly, when both $T'((x)_R, t_1)$ and $T'((x)_R, t_1 + t_2)$ are defined, we obtain the same equality.

The dynamics T' is called the *quotient dynamics* on S' induced by T.

REMARK. Note that the equation $T'((x)_R, t) = (T(x,t))_R$ states that the diagram

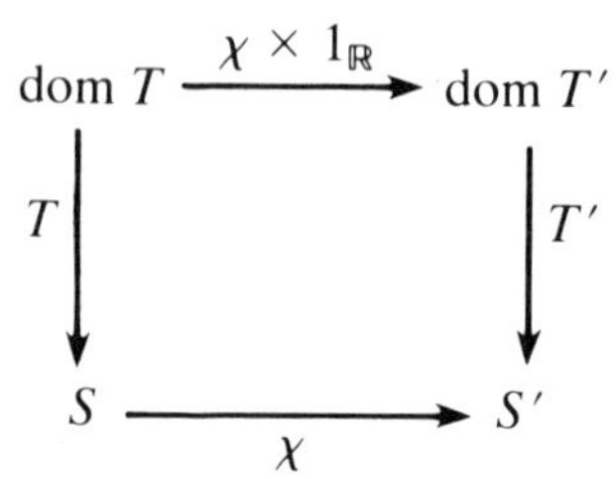

commutes (where $\chi: S \to S'$ is the quotient map $x \mapsto (x)_R$). So $\chi: (S, \{T\}) \to (S', \{T'\})$ sending each $x \in S$ to $(x)_R \in S'$ and sending T to T' is a **D**-morphism (in fact, a **D**-epimorphism).

19. Bijections and Equivalence Relations

There are two complementary questions on the connection between dynamics on S and equivalence relations on S that can be asked:

(1) Given a dynamics T on S, how can we characterize those equivalence relations R on S with which T is compatible?

(2) Given an equivalence relation R on S, how can we characterize those dynamics on S that are compatible with R?

The simpler case when one considers bijections ('automorphisms') $T: S \to S$ instead of dynamics was discussed in Rosen (1978) [in which a bijection $T: S \to S$ is defined to be compatible with an equivalence relation R on S if sRs' implies $(Ts)R(Ts')$]. It was shown that the set of all equivalence relations with which a given bijection $T: S \to S$ is compatible forms a sublattice of the lattice $\mathscr{R}(S)$ of all equivalence relations on S and that the set of all bijections compatible with a given equivalence relation forms a submonoid of the group $\mathscr{A}(S)$ of all bijections on S; further, the set of all bijections compatible with R, and whose inverses are also compatible with R, forms a subgroup of $\mathscr{A}(S)$. [Note that when both T and T^{-1} are compatible with R, sRs' iff $(Ts)R(Ts')$.]

Let us first expand on these ideas.

20. Galois Theory

DEFINITION. A subset $\mathscr{A}$ of $\mathscr{A}(S)$ is *compatible* with a family $\mathscr{R}$ of equivalence relations on S if for each $T \in \mathscr{A}$, for each $R \in \mathscr{R}$, and for all $s, s' \in S$, sRs' implies $(Ts)R(Ts')$.

We shall now try to answer these two questions:

(1*) Given a subgroup G of $\mathscr{A}(S)$, what are the equivalence relations on S with which G is compatible?

(2*) Given a sublattice L of $\mathscr{R}(S)$, what are the bijections on S such that they and their inverses are compatible with L?

The same argument used to obtain the results of Section IV.19 can easily be generalized to answer these questions. The answer to (1*) is a sublattice of $\mathscr{R}(S)$ which contains the equality relation $=$ and the "universal" relation S^2 defined by sS^2s' iff $s, s' \in S$. Since every bijection is compatible with $=$ and S^2, it is without loss of generality to (and hence we shall) only consider sublattices that contain both of these relations. The answer to (2*) is a subgroup of $\mathscr{A}(S)$. In fact, this correspondence between the set of all sublattices of $\mathscr{R}(S)$ and the set of all subgroups of $\mathscr{A}(S)$ turns out to be bijective and order (*qua* substructures)-inverting. We can represent this situation by the following self-explanatory diagram:

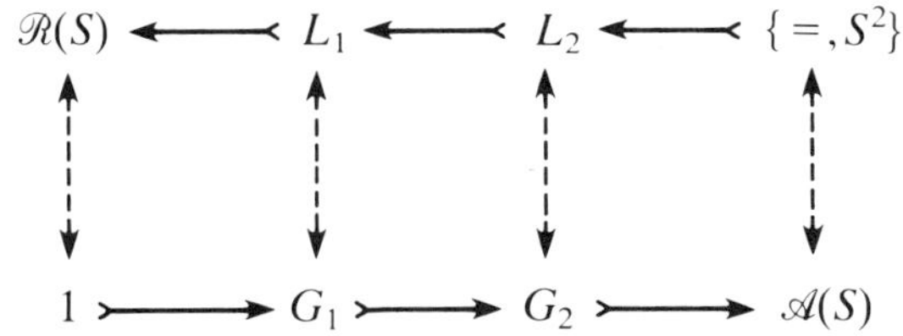

Further, we can define that an equivalence relation R' on S is *conjugate* to another equivalence relation R on S if there is a bijection $T: S \to S$ such that $sR's'$ if and only if $(Ts)R(Ts')$, in which case we shall write $R' = R_T$. Conjugacy is equivalent to the existence of an isomorphism between the collections of equivalence classes S/R and S/R', i.e., there is a bijection between the equivalence classes in S/R and S/R' and that corresponding equivalence classes contain the same number of S-elements.

Let L be a sublattice of $\mathscr{R}(S)$, then we can define a *conjugate* of L to be $L_T = \{R_T : R \in L\}$ for some $T \in \mathscr{A}(S)$. With these definitions we have the following interesting lemma.

LEMMA. If (L, G) is a pair of corresponding sublattice-subgroup with respect to compatibility, then the conjugate $(L_T, T^{-1}GT)$ is also a pair of corresponding sublattice-subgroup with respect to compatibility.

Proof. Let $R \in L$ and $U \in G$, then for $s, s' \in S$,

$$\begin{aligned} sR_Ts' \quad &\text{iff} \quad (Ts)R(Ts') && [\text{definition of } R_T] \\ &\text{iff} \quad (UTs)R(UTs') && [(R,U) \in (L,G)] \\ &\text{iff} \quad (TT^{-1}UTs)R(TT^{-1}UTs') && [TT^{-1} = 1_S] \\ &\text{iff} \quad (T^{-1}UTs)R_T(T^{-1}UTs') && [\text{definition of } R_T] \end{aligned}$$

Thus $T^{-1}UT$ and $(T^{-1}UT)^{-1}$ are both compatible with R_T. □

The preceding results bear a striking resemblence to the Galois theory of field extensions and automorphism groups!

21. Discrete Dynamical Systems

Before going further, let us take a digression in the following direction. Suppose instead of considering time as the continuum of real numbers we consider time as being composed of a succession of "elementary steps." Then we can define a *discrete dynamics* on a set S to be a mapping T from $S \times \mathbb{Z}$ to S such that

(i) $T(x, 0) = x$ for all $x \in S$, and
(ii) $T(T(x, t_1), t_2) = T(x, t_1 + t_2)$ for all $x \in S$ and all $t_1, t_2 \in \mathbb{Z}$.

For simplicity we have assumed that dom T is all of $S \times \mathbb{Z}$ instead of a subset as in Definition IV.3 of a "continual" dynamics.

It is clear that for each $n \in \mathbb{Z}$, each $T(\cdot, n)$ is a bijection from S to S and that $T(\cdot, n) = T(\cdot, 1)^n$. Thus, alternatively, we can define a discrete dynamics to be the cyclic subgroup of $\mathscr{A}(S)$ generated by a bijection $T \in \mathscr{A}(S)$, i.e., a discrete dynamics is $\langle T \rangle = \{T^n : n \in \mathbb{Z}\} \subset \mathscr{A}(S)$. With this terminology the powers of

T in $\langle T \rangle$ become interpretable as *instants of logical time*, and the transition $x \mapsto Tx$, or $T^{n-1}x \mapsto T^n x$ in general, is an *elementary step* of the dynamics. Further, with appropriate modifications, all the discussions on the category **D** still go through with $\mathbb{R}$ replaced by $\mathbb{Z}$ and we would have the category of *discrete* dynamical systems instead.

This situation is, of course, closely related to the specification of a continual dynamical system as a one-parameter group of bijections $T = \{T_t : t \in \mathbb{R}\}$ on a phase space S. (Let us again restrict to the case when dom $T = S \times \mathbb{R}$, i.e., when dom $T_t = S$ for all t.) For any real number r, we can consider the cyclic subgroup generated by T_r, then $\langle T_r \rangle = \{T_r^n = T_{nr} : n \in \mathbb{Z}\}$ defines a discrete dynamics on S. This method of "discretization" is used, for example, in obtaining numerical solutions of differential equations. Note, however, that this procedure only goes one way: We can obtain a discrete time from a continuous time by choosing the size of an elementary step $t = r$, but starting from a discrete dynamics $\langle T \rangle$, in general we cannot embed it into a continuous one-parameter group of bijections. A further discussion of this aspect can be found in Rosen (1981). We shall come back to the role of time in dynamics at the end of Section V.

22. *Discrete Dynamics and Equivalence Relations*

Now let us try to answer the two questions in Section IV.19 for discrete dynamics.

Given a discrete dynamics $\langle T \rangle = \{T^n : n \in \mathbb{Z}\}$, $T \in \mathscr{A}(S)$, since $\langle T \rangle$ is in particular a (cyclic) subgroup of $\mathscr{A}(S)$, the results of Section IV.20 tell us that the set of all equivalence relations with which $\langle T \rangle$ is compatible forms a sublattice of $\mathscr{R}(S)$. It is easy to see that $\langle T \rangle$ is compatible with an equivalence relation R on S if and only if both T and T^{-1} are compatible with R, so we only need to check the compatibility of the generator and its inverse with R.

Since given an equivalence relation R on S, the set of all bijections on S compatible with R and whose inverses are also compatible with R forms a subgroup G of $\mathscr{A}(S)$, the set of all discrete dynamics on S compatible with R is the collection of all cyclic subgroups of G, i.e., the collection of all homomorphic images of $(\mathbb{Z}, +)$ in G.

23. *Partial Answers*

Finally, we shall try to answer, at least partially, the two questions in Section IV.19. We shall only consider the dynamics T for which dom $T = S \times \mathbb{R}$.

Given a dynamics T on S, the collection of translations $\{T_t : t \in \mathbb{R}\}$ is a subgroup of $\mathscr{A}(S)$; thus the set of all equivalence relations with which T is compatible is a sublattice of $\mathscr{R}(S)$.

The second question is more difficult. Given an equivalence relation R on S, we obtain a subgroup G of $\mathscr{A}(S)$ of all bijections on S compatible with R. Now we have to find subgroups of G that are (isomorphic to) continuous one-parameter groups indexed by $\mathbb{R}$. Whereas for the discrete dynamics case we can get the set $\{\{T^n : n \in \mathbb{Z}\} : T \in G\}$ quite easily, here there is no trivial way to look for homomorphic images of $(\mathbb{R}, +)$ in G. Note, however, that there is at least one dynamics compatible with R, namely, the trivial dynamics $I_S = \{1_S : t \in \mathbb{R}\} \subset G$.

V. TOPOLOGICAL DYNAMICS

That which gives things their suchness
Cannot be delimited by things.
So when we speak of "limits" we remained confined
To limited things.

Chuang Tse

In this section the study of dynamical systems is continued. First, in Subsections A and B, we shall consider the topology on the phase space induced by a dynamics; then we shall consider an arbitrary topology on the phase space of a dynamics such that the dynamics is a continuous map. The development in these sections is adapted from Hájek (1968). Next, via topological constructions, the relations between observables and dynamics on the same set are analysed in Subsection C. Finally, in Subsection D, the role of time in dynamics is discussed.

A. *T*-Topology

Let T be a dynamics on a set S.

DEFINITION. A subset $G \subset S$ is *T-open* if for every $x \in G$ there exists an $\varepsilon > 0$ such that $T(x, (-\varepsilon, \varepsilon)) = \{T(x,t): -\varepsilon < t < \varepsilon\} \subset G$.

The collection of T-open sets clearly forms a topology on S with the collection $\{T(x, (-\varepsilon, \varepsilon)) : x \in S, \varepsilon > 0\}$ as a base. This topology is called the *T-topology* on S.

1. Properties

Some immediate consequences of the definition are:

(i) A net $\{x_i\}$ is *T-convergent* to x in S if and only if for large i, $x_i = T(x, t_i)$ with $t_i \to 0$ in $\mathbb{R}$. This also characterizes *T-closed* sets.

(ii) S is *T-locally compact.*

(iii) S is *T-locally pathwise connected.*

(iv) A subset of S is *T-invariant* if and only if it is *T-clopen* (closed and open). So all trajectories of T are T-clopen (see Section IV.7).

(v) S is *T-connected* if and only if it consists of a single trajectory (or if S is empty).

(vi) If S is *T-compact*, then it consists of a finite set of trajectories.

LEMMA 1. $\operatorname{dom} T \subset S \times \mathbb{R}$ is T-open. The map $T: \operatorname{dom} T \to S$ is T-continuous. (The topologies on the various sets are, naturally, the T-topology for S, the usual topology for $\mathbb{R}$, the product topology for $S \times \mathbb{R}$, and the subspace topology for $\operatorname{dom} T \subset S \times \mathbb{R}$.)

Proof. It suffices to show that the inverse image under T of the basic T-open set $G = T(x, (-\varepsilon, \varepsilon))$ is open in $S \times \mathbb{R}$. Now if $(x', t') \in T^{-1}(G)$, then $T(x', t') = T(x, t)$ with $|t| < \varepsilon$. Let

$$\delta = \tfrac{1}{2}(\varepsilon - |t|)$$

and consider $E = T(x', (-\delta, \delta)) \times (t' - \delta, t' + \delta)$. E is then clearly an open neighbourhood of (x', t') in $S \times \mathbb{R}$ with $T(E) = T(T(x', t'), (-2\delta, 2\delta)) = T(T(x, t), (|t| - \varepsilon, \varepsilon - |t|)) = T(x, (t + |t| - \varepsilon, t - |t| + \varepsilon)) \subset T(x, (-\varepsilon, \varepsilon)) = G$. Thus $T^{-1}(G)$ is open in $S \times \mathbb{R}$. □

LEMMA 2. The T-topology is the finest among topologies on S rendering all solutions $y_x: (a_x, b_x) \to S$ continuous.

Proof. It follows from Lemma 1 that all y_x are T-continuous. On the other hand, let τ be a topology on S in which all y_x are continuous. Let $G \in \tau$ and $x \in G$. Then $y_x(0) = x \in G$ so $T(x, (-\varepsilon, \varepsilon)) = y_x(-\varepsilon, \varepsilon) \subset G$ for small $\varepsilon > 0$. Thus G is open in the T-topology, which is therefore finer than τ. □

THEOREM 3. The T-topology is the finest among topologies on S rendering $T: \operatorname{dom} T \to S$ continuous. □

COROLLARY. The bounds a_x and b_x, considered as maps from S to $\bar{\mathbb{R}}$, are T-continuous.

Proof. Theorem 3 and Lemma IV.6. □

LEMMA 4. Let $A \subset S$. Then T induces a relative dynamics T' on A if and only if A is T-open in S. In the positive case, the T'-topology on A coincides with the subspace T-topology of A in S.

Proof. $\Rightarrow$ If $x \in A$, then by the definition of dynamics, $T'(x, t) \in A$ for small $|t|$. Since T' is the restriction of T, we have $T(x, t) \in A$ for small $|t|$, so A is T-open.

$\Leftarrow$ Suppose A is T-open. Let $x \in A$. For $t \geq 0$ define $T'(x,t) = T(x,t)$ if $T(x,t') \in A$ for all $t' \in [0,t]$. Similarly, for $t \leq 0$. Otherwise leave $T'(x,t)$ undefined. Then clearly T' is the relative dynamics on A induced by T. The bounds $a_x[T']$ and $b_x[T']$ are determined by

$$a_x[T'] = \inf\{t : T(x,t') \in A \text{ for } t \leq t' \leq 0\},$$

$$b_x[T'] = \sup\{t : T(x,t') \in A \text{ for } 0 \leq t' \leq t\}.$$

It is clear that the two topologies on A coincide. □

Compare IV.7, property (iv), and this lemma:

$$\begin{array}{ccc} A \text{ is } T\text{-invariant} & \overset{\text{(iv)}}{\Longleftrightarrow} & A \text{ is } T\text{-clopen} \\ \text{IV.7} \Big\Downarrow & & \Big\Downarrow \\ T \text{ induces a relative dynamics on } A & \underset{\text{Lemma 4}}{\Longleftrightarrow} & A \text{ is } T\text{-open} \end{array}$$

LEMMA 5. Let $\phi:(2_1, D_1) \to (S_2, D_2)$ be a **D**-morphism. Then for each $T \in D_1$, $\phi: S_1 \to S_2$ is a continuous function where S_1 and S_2 are endowed with the T- and ϕT-topology, respectively.

Proof. It follows from Lemma 2 that $\phi: S_1 \to S_2$ is continuous if and only if $\phi \circ y_x: \mathbb{R} \to S_1 \to S_2$ is continuous for each solution y_x of T. By definition of a **D**-morphism $\phi(T(x,t)) = \phi T(\phi x, t)$, hence $\phi(y_x(t)) = y'_{\phi x}(t)$ (where $y'_{x'}$ denotes solutions of ϕT), and $a_{\phi x}[\phi T] \leq a_x[T] < 0 < b_x[T] \leq b_{\phi x}[\phi T]$ (see Definition IV.9). Thus $\operatorname{dom}(\phi \circ y_x) = (a_x[T], b_x[T])$ is an open (in $\mathbb{R}$) subset of $\operatorname{dom} y'_{\phi x} = (a_{\phi x}[\phi T], b_{\phi x}[\phi T])$, and so $\phi \circ y_x$ is continuous because $y'_{\phi x}$ is. □

B. Continuous Dynamics

2. *Definition and Consequences*

DEFINITION. Let T be a dynamics on S and τ be a topology on S. T is a *continuous dynamics* on (S, τ) (or simply "on S" if τ is clear or immaterial) if

(i) $T: \operatorname{dom} T \to S$ is continuous, and
(ii) $\operatorname{dom} T$ is open in $S \times \mathbb{R}$

(where, of course, $S \times \mathbb{R}$ has the product topology of τ on S and the usual topology on $\mathbb{R}$).

Note condition (i) implies that τ is coarser than the T-topology on S [see Theorem V.1(3)]. If $\operatorname{dom} T = S \times \mathbb{R}$, then (ii) is automatically satisfied so a continuous dynamics in this case is simply a dynamics that is continuous.

Conditions (i) and (ii) are equivalent to each of the following:

(a) If the net $\{x_i\}$ converges to x in S, the net $\{t_i\}$ converges to t in $\mathbb{R}$, and $T(x, t)$ is defined, then $T(x_i, t_i) \to T(x, t)$.

(b) The inverse image under T of an open set in S is open in $S \times \mathbb{R}$.

EXAMPLES. (1) T is a continuous dynamics on S with the T-topology [Lemma V.1(1)].

(2) T is a continuous dynamics on S with the discrete topology if and only if $T = I_S =$ the trivial dynamics on S. In fact, the discrete topology is the I_S-topology on S.

(3) The trivial dynamics on S is a continuous dynamics on S with any topology.

(4) T is a continuous dynamics on S with the indiscrete topology $(= \{S, \varnothing\})$ if and only if $\operatorname{dom} T = S \times \mathbb{R}$.

(5) (See Section IV.4) The dynamics associated with an autonomous differential equation $dx/dt = f(x)$ on an open subset $S \subset \mathbb{R}^k$, where $f: S \to \mathbb{R}^k$ is as in Section IV.4, is a continuous dynamics on S with the usual topology. Note that condition (i) corresponds to continuous dependence on initial data.

LEMMA 1. If T is a continuous dynamics on S, then the bounds a_x and b_x are, respectively, upper and lower semicontinuous maps $S \to \mathbb{R}$; i.e., $a_x \geq \limsup_{y \to x} a_y$ and $b_x \leq \liminf_{y \to x} b_y$.

Proof. Let $x \in S$ and $t \in (a_x, 0]$. Then $(x, t) \in \operatorname{dom} T$, so there is a neighbourhood N around x such that for $y \in N$, $(y, t) \in \operatorname{dom} T$ (since $\operatorname{dom} T$ is open in $S \times \mathbb{R}$). Hence $t \in (a_y, 0]$ for these y, and so $t \geq \limsup_{y \to x} a_y$. Now take $t \to a_x$.

Similarly for b_x. ☐

COROLLARY. If T is a continuous dynamics on S, then for any $t \in \mathbb{R}$, the sets $\{x \in S : a_x < t\}$ and $\{x \in S : t < b_x\}$ are open in S, so that $\{x \in S : a_x = -\infty\}$ and $\{x \in S : b_x = +\infty\}$ are G_δ's in S. ☐

LEMMA 2. If T is a continuous dynamics on S, then each solution y_x is continuous, and each translation T_t is continuous with domain open in S. ☐

LEMMA 3. If T is a continuous dynamics on (S, τ) and $A \subset S$ is either τ-open or T-invariant, then the relative dynamics on A induced by T is a continuous dynamics on A with the subspace τ-topology.

Proof. It follows from Theorem V.1(3) and property V.1(iv) that in either case A is T-open, so Lemma V.1(4) gives this result. □

REMARK. Lemmata V.1(4) and V.2(3) give connections between continuity and **D**-subsystems. Lemma V.1(5) gives connections between continuity and **D**-morphisms. With Theorem V.1(3), this says that when $\phi:(S_1, D_1) \to (S_2, D_2)$ is a **D**-morphism, and if ϕT is a continuous dynamics on S_2 with some topology τ, then $\phi: S_1 \to S_2$ is continuous where S_1 has the T-topology and S_2 has the topology τ. Note, however, this does *not* imply that if T is a continuous dynamics on (S_1, τ'), then $\phi: S_1 \to S_2$ is continuous with respect to τ' and the ϕT-topology; the implication only goes one way.

Now how about the connections between continuity and **D**-products, **D**-coproducts, and quotient dynamics? The answers are given in the next few theorems.

THEOREM 4. For each $i \in I$ let T_i be a continuous dynamics on (S_i, τ_i). If the product dynamics $T = (T_i: i \in I)$ on $S = \Pi\, S_i$ exists in **D**, then T is a continuous dynamics on (S, τ) where τ is the product topology of the τ_i's.

Proof. By Theorem IV.12 there is a finite set $J \subset 1$ such that dom $T_i = S_i \times \mathbb{R}$ for $i \notin J$.

To show that T is a continuous dynamics on (S, τ), it is necessary and sufficient to show that the inverse image of a τ-open set $G \subset S$ under T is open in $S \times \mathbb{R}$ [see condition (b)]. It suffices to let G be a subbasic open set $G = G_k \times \Pi_{i \neq k} S_i$ for some $k \in I$, G_k open in S_k. Also, without loss of generality assume $k \notin J$.

Let $x = (x_i: i \in I) \in S$, $t \in \mathbb{R}$, and $T(x, t) \in G$. Then in particular $T_k(x_k, t) \in G_k$ so condition (b) applied to T_k yields a τ_k-open $H_k \subset S_k$, open $E_k \subset \mathbb{R}$, with $(x_k, t) \in H_k \times E_k \subset T_k^{-1}(G_k) \subset \text{dom}\, T_k$. Doing the same thing for each $i \in J$, we obtain two families of open sets $\{H_i: i \in J\}$ and $\{E_i: i \in J\}$. Let

$$H = H_k \times \prod_{i \in J} H_i \times \prod_{\substack{i \notin J \\ i \neq k}} S_i \subset S,$$

$$E = E_k \bigcap \bigcap_{i \in J} E_i \subset \mathbb{R}.$$

Then $H \times E$ is open in $S \times \mathbb{R}$ and $(x, t) \in H \times E \subset T^{-1}(G) \subset \text{dom}\, T$, hence $T^{-1}(G)$ is open. □

THEOREM 5. Let T_i be a continuous dynamics on (S_i, τ_i) for each $i \in I$. Then each of the coproduct dynamics (i, T_i) in **D** is a continuous dynamics on $S = \amalg\, S_i$ with the direct sum topology.

Proof. (See Section IV.13) The proof is immediate; recall that a set $G \subset S$ is open relative to the direct sum topology if and only if for each $i \in I$, $G \cap S_i$ is τ_i-open. □

THEOREM 6. Let T be a continuous dynamics on (S, τ) and let T be compatible with an equivalence relation R on S. Let T' be the quotient dynamics induced by T on $S' = S/R$, and let τ' be the quotient topology on S'. Then T' is a continuous dynamics on (S', τ').

Proof. (See Section IV.18) The equivalence relation R on S induces an equivalence relation R on $S \times \mathbb{R}$ via $(x, t)R(x', t')$ if and only if xRx' and $t = t'$. Then $(S \times \mathbb{R})/R \cong S' \times \mathbb{R}$. Also, since the bounds are such that $a_{(x)_R}[T'] = a_x[T]$ and $b_{(x)_R}[T'] = b_x[T]$, we have $\text{dom } T/R \cong \text{dom } T'$, and the topology on dom T' is in fact the quotient of the topology on dom T modulo R.

Now consider the commutative diagram where $\chi: S \to S'$ is the natural projection, and hence $\chi \times 1_{\mathbb{R}}: \text{dom } T \twoheadrightarrow \text{dom } T' = \text{dom } T/R$ is also the corresponding natural projection:

$$\begin{array}{ccc} \text{dom } T & \xrightarrow{\chi \times 1_{\mathbb{R}}} & \text{dom } T' \\ {\scriptstyle T}\downarrow & & \downarrow{\scriptstyle T'} \\ S & \xrightarrow[\chi]{} & S' \end{array}$$

Let $G \subset S'$ be τ'-open. Since both T and χ are continuous, $(\chi \circ T)^{-1}(G) \subset \text{dom } T$ is open (in the product topology of $S \times \mathbb{R}$). But $(\chi \circ T)^{-1}(G) = (\chi \times 1_{\mathbb{R}})^{-1} \circ (T')^{-1}(G)$, so by the definition of the quotient topology on dom T', $(T')^{-1}(G)$ must be open (in the topology of $S' \times \mathbb{R}$). Thus T' is a continuous dynamics on (S', τ'). □

C. Observables and Induced Topologies

3. *Observables Revisited*

Let f be an observable on the set of states S, i.e., $f: S \to \mathbb{R}$. In the construction of the category **S**, we could have used equivalence classes of functions in $\mathbb{R}^S/\sim$ as observables (see Section III.2) because there the only relevant property of f was the equivalence relation R_f it imposed on S. But such a definition of **S**-objects apparently leads to difficulties in some categorical constructions, and creates problems in topological considerations.

But we should recall the comments in Section III.7 and note that for $f, g \in \mathbb{R}^S$ and $f \sim g$ (i.e., $R_f = R_g$), $(S, \{f\})$ and $(S, \{g\})$ are isomorphic in **S**, and because all constructions in a category are only "up to isomorphism," these two **S**-objects (and any constructions with one or the other) are "indistinguishable" in **S**.

4. Induced Topology

What properties of $f \in \mathbb{R}^S$ are hidden if we consider $(f)_\sim \in \mathbb{R}^S/\sim$ instead? We see that since $\mathbb{R}$ is a topological space (with the usual topology), $f: S \to \mathbb{R}$ can induce a topology on S, called the *f-topology*, as follows. A subset of S is *f-open* (respectively, *f-closed*) if and only if it is the inverse image under f of an open (respectively, closed) subset of $\mathbb{R}$. The f-topology is the coarsest topology on S that renders f continuous.

For any constant function f [i.e., for $f \in (0)_\sim \in \mathbb{R}^S/\sim$], the f-topology is the indiscrete topology $\{\varnothing, S\}$ on S. And in general if the range of f is a finite set, then any representative of the class $(f)_\sim$ induces the same f-topology on S, and so there is a unique $(f)_\sim$-topology. But if the range of f is infinite, then it is possible for $f(S)$ to have limit points, in which case different class representatives of $(f)_\sim$ may induce different topologies; hence one has to consider $f \in \mathbb{R}^S$ and not $(f)_\sim \in \mathbb{R}^S/\sim$ in topological considerations.

Further, since $\mathbb{R}$ is a metric space, a "distance function" on S can be defined using f. Namely, for $x, y \in S$, define $d_f(x, y) = |f(x) - f(y)|$. It is clear that $d_f(x, x) = 0$, that $0 \leq d_f(x, y) = d_f(y, x) < +\infty$, and that d_f satisfies the triangle inequality. But $d_f(x, y) = 0$ only means $f(x) = f(y)$ and not necessarily $x = y$, so d_f is a *pseudometric* on S, and d_f is a metric if and only if f is injective. Obviously, the f-topology on S is the pseudometric topology generated by d_f, and the quotient f-topology on S/R_f is the induced metric topology. Note that if $x \in G \subset S$ where G is f-open and $d_f(x, y) = 0$, then $y \in G$.

Historically, the ideas of limit and continuity appeared very early in mathematics, notably in geometry, and their role has steadily increased with the development of analysis and its applications to the experimental sciences, since these ideas are closely related to those of *experimental determination* and *approximation*. But since most experimental determinations are *measurements*, i.e., determinations of one or more numbers, it is hardly surprising that the notions of limit and continuity in mathematics were featured at first only in the theory of real numbers and its outgrowths and fields of application. So in a sense topology has its roots in the process of measurement (i.e., observations) and it is interesting to note that we are now using topology as a tool in the study of the fundamentals of measurement and representation of natural systems.

5. *f*-Continuity

DEFINITION. Let S be a set, T a dynamics on S, and $f \in \mathbb{R}^S$ an observable. Then T is *f-continuous* if T is a continuous dynamics on S with the f-topology (whence the f-topology is coarser than the T-topology).

6. Compatibility

Let T be a dynamics on S and R an equivalence relation on S. Recall (Definition IV.17) that T is *compatible with* R if xRy implies $T(x, t)$ is defined if and only if $T(y, t)$ is defined, and $T(x, t)RT(y, t)$; i.e., for all $t \in \mathbb{R}$ and for all $x \in \operatorname{dom} T_t \subset S$, xRy implies $y \in \operatorname{dom} T_t$ and $T_t(x)RT_t(y)$.

In particular, if $f \in \mathbb{R}^S$, then T is compatible with R_f (or simply T is *compatible with* f) if for all $t \in \mathbb{R}$ and for all $x \in \operatorname{dom} T_t$, $f(x) = f(y)$ implies $y \in \operatorname{dom} T_t$ and $f(T_t(x)) = f(T_t(y))$.

THEOREM 1. Let T be a dynamics on S and $f \in \mathbb{R}^S$. If T is f-continuous, then T is compatible with f.

Proof. If T is f-continuous, then each T_t is continuous on $\operatorname{dom} T_t \subset S$ with the subspace f-topology and $\operatorname{dom} T_t$ is f-open (see Lemma V.2(2)). This means that for all $x \in \operatorname{dom} T_t$ and for all $\varepsilon > 0$ there exists a $\delta > 0$ such that

$$d_f(x, y) = |f(x) - f(y)| < \delta$$

implies $y \in \operatorname{dom} T_t$ and $d_f(T_t(x), T_t(y)) = |f(T_t(x)) - f(T_t(y))| < \varepsilon$. Now let $x \in \operatorname{dom} T_t$ and $y \in S$ be such that $f(x) = f(y)$. Then for every $\varepsilon > 0$, $d_f(x, y) = |f(x) - f(y)| = 0 < \delta(\varepsilon)$, hence $y \in \operatorname{dom} T_t$ and $d_f(T_t(x), T_t(y)) = |f(T_t(x)) - f(T_t(y))| < \varepsilon$; so $|f(T_t(x)) - f(T_t(y))| = 0$, whence $f(T_t(x)) = f(T_t(y))$. Thus T is compatible with f. □

It is clear that the converse of this theorem is not necessarily true. Further, combining Theorems V.2(6) and V.6(1), we have the following theorem.

THEOREM 2. Let T be a dynamics on S and $f \in \mathbb{R}^S$, and let T be f-continuous. Then the quotient dynamics T' induced by T on the metric space $(S/R_f, d_f)$ is a continuous dynamics. □

7. *T*-Continuity

DEFINITION. Let T be a dynamics on S and $f \in \mathbb{R}^S$. Then f is *T-continuous* if $f\colon S \to \mathbb{R}$ is a continuous function on S with the T-topology (and $\mathbb{R}$ with the usual topology).

Since the f-topology is the coarsest topology on S rendering f continuous, f is T-continuous if and only if the T-topology is finer than the f-topology.

Trivially, we have

LEMMA. (i) $\Rightarrow$ (ii) $\Leftrightarrow$ (iii), where

(i) T is f-continuous.
(ii) The T-topology is finer than the f-topology.
(iii) f is T-continuous. □

D. Time and Dynamics

8. *Time as Real Numbers*

A main area of investigation in this study is to determine how we may employ formal dynamical models of natural systems to make spatial and temporal predictions about the systems themselves. So let us examine the modelling relations involving dynamical systems in some detail.

The crucial concept in dynamics is, of course, *time*. The concept of time involves two distinct aspects: *simultaneity* and *temporal succession*. Both of these are intimately involved in our study of dynamical systems. In our definition of dynamics, we have tacitly encoded time as a "set of instants" in the continuum $\mathbb{R}$, and we have made extensive use of the mathematical properties of $\mathbb{R}$. Simultaneity and temporal succession are implied by the total order $\leq$ on $\mathbb{R}$. The fact that $(\mathbb{R}, +)$ is an abelian group is basic to the group property of dynamics [see Definition IV.3(iii)].

In this chapter when we looked at topological dynamics, the topological (metric) properties of $\mathbb{R}$ entered in an essential way. This is precisely where the difference between a "continual dynamical system" $\{T_t : t \in \mathbb{R}\}$ and a "discrete dynamical system" $\{T^n : n \in \mathbb{Z}\}$ lies (cf. Section IV.21). Although $(\mathbb{Z}, \leq, +)$ is also a totally ordered abelian group, time encoded as real numbers has the further bonus of the expression of "temporal approximation" over the encoding as integers.

Note, however, that this distinction between $\mathbb{R}$ and $\mathbb{Z}$ is purely mathematical: that $\mathbb{R}$ is "complete" while $\mathbb{Z}$ has the discrete topology. When it comes down to tackling fundamental problems like "What is time?" and "What is a universal encoding for time?" it may very well turn out that $\mathbb{Z}$ suffices. After all, since we are only dealing with operational definitions (see Section III.1), we cannot in principle (and obviously in practice as well) measure any duration of time that is shorter than the time required for light to pass through a single hydrogen atom. So if we take this (very small, but nonetheless positive) theoretical lower bound as our elementary step of time, then any period of time becomes an integral multiple of this elementary step, and hence it suffices to encode time as $\mathbb{Z}$. But because we need the topological properties of the continuum, we shall not dwell on this point and simply encode time as $\mathbb{R}$.

9. Dynamical Time

Let us revisit the rate equations we considered in Section IV.4 and Example V.2(5). But this time we shall consider two sets of them:

$$dx/dt_1 = f(x), \qquad dy/dt_2 = g(y), \tag{1}$$

where x is in an open subset S_1 of $\mathbb{R}^m$, y is in an open subset S_2 of $\mathbb{R}^n$ (m and n may be different), $f: S_1 \to \mathbb{R}^m$, $g: S_2 \to \mathbb{R}^n$ are continuous, and t_1 and t_2 are "times."

In employing rate equations of the form (1) as encodings of dynamical processes in natural systems, the usual point of view is that, roughly, the differential increments $dx = (dx_1, dx_2, \ldots, dx_m)$ of the state variables are known, and the time differential dt_1 is known, and $f = (f_1, f_2, \ldots, f_m)$ is their ratio. We can take a different point of view and say that the principle of causality (see Section IV.5) together with the differentials dx essentially serve to specify the observables f (see Section IV.2: a "change of state" is an observable). Observables f and differentials dx then together *define* the time differential dt_1. Similarly, for the other system, g and dy together define the time differential dt_2. One expects that the differentials arising from different dynamical systems are in general quite different from one another, and hence there is no reason to believe that the time differentials dt_1 and dt_2 are the same, i.e., every dynamics defines its own (intrinsic) time. So the problem is how these different times are related to a common 'clock time' (extrinsic), which is likely to be different from all of them.

To tackle this problem let us reconsider equations (1) and this time take the point of view that t is simply an arbitrary parameter; for example, we could take it to be arc length along the trajectories. Then we can multiply the equations by some nonvanishing function $a_i: S_i \to \mathbb{R}$ without changing the qualitative properties of the systems. All we do is change the rates at which the trajectories are traversed (relative to some fixed extrinsic time scale). In effect, we are replacing dt_i by $dt = a_i(\cdot)dt_i$. It is through this 'scaling' that we convert different intrinsic times to a 'common time' in terms of which dynamical predictions can be made.

So when we define $\phi \in \mathbf{D}((S_1, D_1), (S_2, D_2))$ via the diagram ($T \in D_1$):

$$\begin{array}{ccc} \operatorname{dom} T & \xrightarrow{\phi \times 1_{\mathbb{R}}} & \operatorname{dom} \phi T \\ {\scriptstyle T}\downarrow & & \downarrow{\scriptstyle \phi T} \\ S_1 & \xrightarrow[\phi]{} & S_2 \end{array}$$

the $1_{\mathbb{R}}: \mathbb{R} \to \mathbb{R}$ that appears is really a map that makes the correspondence $t_1 \mapsto t_2$, via the scaling above to a common time t ($dt = a_1(x)\,dt_1 = a_2(y)\,dt_2$); i.e., the element-chasing diagram

$$\begin{array}{ccc} (x,t) & \longmapsto & (\phi x, t) \\ \downarrow & & \downarrow \\ T(x,t) & \longmapsto & \phi(T(x,t)) = \phi T(\phi x, t) \end{array}$$

should really read

$$\begin{array}{ccc} (x,t_1) & \longmapsto & (\phi x, t_2) \\ \downarrow & & \downarrow \\ T(x,t_1) & \longmapsto & \phi(T(x,t_1)) = \phi T(\phi x, t_2) \end{array}$$

But the scaling correspondence $t_1 \leftrightarrow t_2$ is assumed from the very beginning to simplify matters. For further discussions on time scaling, see Richardson and Rosen (1979) and Rosen (1981).

VI. THE CATEGORY OF NATURAL SYSTEMS

Nature is the realization of the simplest conceivable mathematical ideas.

Albert Einstein

A. The Modelling Relation

1. *Introduction*

A *natural system*, naturally, is a part of the external world; i.e., it is a member of the entities outside of us, but it can generate under appropriate conditions percepts (sensory impressions) in us. These percepts are then identified with specific properties of the natural system itself.

The adjective *natural* is used to distinguish the system from a "formal" one, which is part of mathematics and hence is a *construct* of our minds. But then one of the primary functions of our minds is to *organize* percepts and establish *relations* among them, which is a *constructive* ability. Consequently, creations of the mind are *imputed* to the external world, and hence we have essentially obtained *formal models of natural systems*. Modelling is in fact a fundamental quality of the mind.

The main objective of our study is to establish relations between the classes of natural and formal systems. The difficulty and challenge in establishing such relations arise from the fact that the two classes are entirely different. A natural system is essentially a bunch of linked qualities coded by the specific percepts that they generate and by the relations that the mind creates to organize them. So a *natural system is never completely known*: We continually learn about such a system as our means of observation and our understanding grow. A formal system, on the other hand, is a creation of our minds, and so we do not learn about a formal system beyond establishing the consequences of our definitions through applications of the usual inferential rules of mathematical logic, and sometimes modifying the initial definitions. The basic task of our study is thus *relating experiments to theory*.

2. Description, Simile, and Metaphor

In *Poetry*, one of the *Six Classics* of Chinese literature, it is stated that there are three major types of figures of speech: *description*, *simile*, and *metaphor*. These three terms, perhaps not too surprisingly, also found their way into the domain of science.

Description in science needs no further explanation; it is the fundamental of experimentation, collection of observable data.

The essential step in our study lies in the exploitation of simile. We are going to force the name of a percept to also be the name of a formal entity, to force the name of a linkage between percepts to also be the name of a relation between mathematical entities, and to force the various temporal relations characteristic of causality in Nature to be synonymous with the dynamical structure of mathematical objects. In short, simile likens these dissimilar things to one another.

Another way to characterize what we are trying to do here is via the three main stages of Bertrand Russell's "scientific process" mentioned in Section I. We seek to *encode* natural systems into formal ones consistently in the sense that the observed phenomena are accounted for, *infer* further consequences from the mathematical structures, and make *predictions* about the natural systems then *verified* when appropriately *decoded*. Diagrammatically, we have

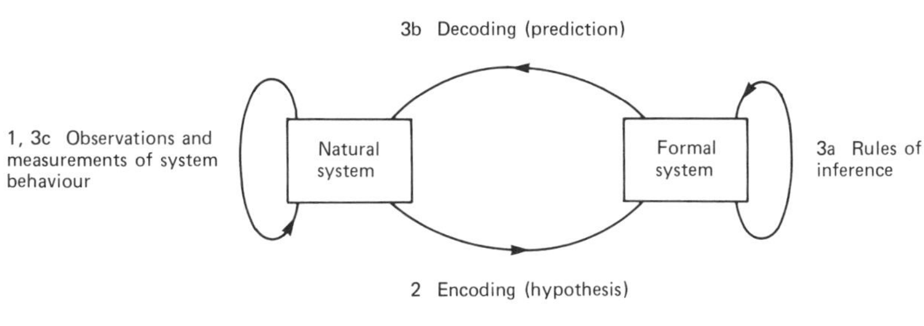

The relation we established between the two preceding systems is the *modelling relation*. We also say that the formal system is a *model* of the natural system, that the latter is a *realization* of the former, and that the two systems are *similes* of each other, via the encoding and decoding rules in question. Note that the idea of simile—the (realization, model) pair—establishes a relation between a natural system and a formal system.

Now suppose we have two natural systems N_1 and N_2 encoded into the same formal system S via encodings E_1 and E_2, respectively:

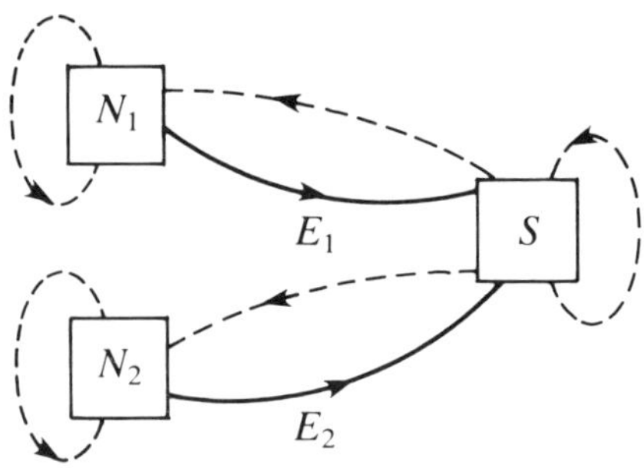

Then a relation between N_1 and N_2 can be established depending on the "degree of overlap" in S between the "images" $E_1(N_1)$ and $E_2(N_2)$. If $E_1(N_1) = E_2(N_2)$, then the natural systems N_1 and N_2 share a common model, or are *analogues* of each other. *If* $E_1(N_1) \subset E_2(N_2)$, then by restricting to a subsystem N'_2 of N_2 we would have $E_1(N_1) = E_2(N'_2)$, so we say N_2 contains a subsystem analogous to N_1. The more general case of $E_1(N_1) \bigcap E_2(N_2) \neq \varnothing$ can be described as N_1 and N_2 possessing subsystems analogous to each other. Note that the relation of analogy is one among natural systems.

Even more generally, we can consider the situation illustrated in the following (self-explanatory) diagram:

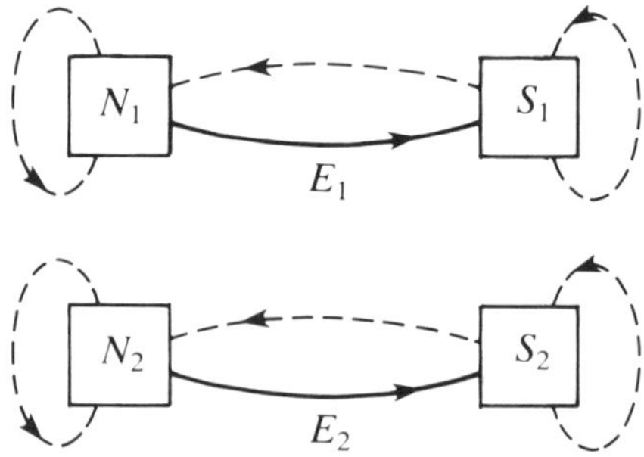

If $E_1(N_1)$ and $E_2(N_2)$ are isomorphic (as mathematical objects in an appropriate category), then via this (not necessarily unique) one-to-one and onto structure-preserving map, we can establish a "dictionary" between the two encodings and hence a relation between N_1 and N_2, called an *extended analogy*. To relax the conditons even further, if $E_1(N_1)$ and $E_2(N_2)$ share some significant general properties in common, a relation can be imputed to N_1 and N_2 themselves; in this case we say that N_1 and N_2 are *metaphors* of each other. Again, the ideas of extended analogy and metaphor are relations among natural systems and can be passed onto subsystems in the same way as in the preceding paragraph.

3. *Observables and Linkages*

Having gone through the preceding discussions, we can now simply say that a natural system is a set of "qualities" on which different definite relations can be imputed. A perceptible quantity of a natural system is obviously what we call an *observable*, and relations among them are *linkages*. The study of natural systems is precisely the specification of its observables, and the characterization of the manner in which they are linked. Thus it becomes clear that the category **S** we looked at in Section III is the appropriate mathematical (formal) tool to be used to study (static models of) natural systems.

Next we have to recognize that (almost by definition) natural systems are dynamic objects and their changes cause a modification in our percepts. Most of the changes in natural systems are of course from their mutual interactions, and in fact the changes in our percepts (these are "observables") can be considered as the result of interactions with other natural systems. So if an interaction between two natural systems causes some change, then the vehicle responsible for the change in one is an observable of the other. This leads us to the discussions of meters and dynamics, and dynamical systems in general, of Section IV. So the category **D** can be used to model the dynamical aspects of natural systems.

We can express these considerations succinctly in a diagram:

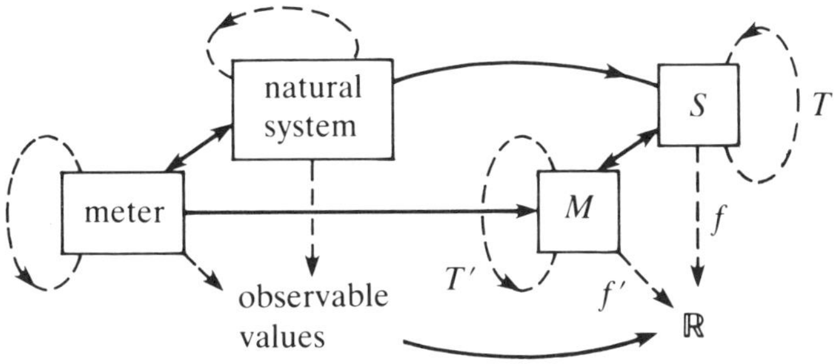

In this section we shall be concerned with these models of natural systems. The categories **S** and **D** will be amalgamated into the "category of natural systems," denoted by **N**.

B. The Category N

4. Products of Categories

From two categories **A** and **B** a new category **A** × **B**, called the *product* of **A** and **B**, can be constructed. An object of **A** × **B** is a pair (A, B), where A is an **A**-object and B a **B**-object; a morphism $(A_1, B_1) \to (A_2, B_2)$ of **A** × **B** is a pair (f, g) of **A**-morphism $f: A_1 \to A_2$ and **B**-morphism $g: B_1 \to B_2$; and the composition of two **A** × **B**-morphisms is defined component-wise by

$$(f_2, g_2) \circ (f_1, g_1) = (f_2 \circ f_1, g_2 \circ g_1).$$

Functors P: **A** × **B** → **A** and Q: **A** × **B** → **B** called *projections* of the product are defined, naturally, by

$$P(A, B) = A, \qquad P(f, g) = f,$$

$$Q(A, B) = B, \qquad Q(f, g) = g.$$

The triple (**A** × **B**, P, Q) clearly satisfies the universal property for categorical products and is in fact the **Cat**-product of **A** and **B**.

5. The Category **S** × **D**

Objects of **S** × **D** look like $((S, F), (S', D))$ and morphisms in **S** × **D** are $(\phi, \phi'): ((S_1, F_1), (S'_1, D_1)) \to ((S_2, F_2), (S'_2, D_2))$, where $\phi \in \mathbf{S}((S_1, F_1), (S_2, F_2))$ and $\phi' \in \mathbf{D}((S'_1, F_1), (S'_2, F_2))$.

A general **S** × **D**-object (in particular when $S \neq S'$) would be of interest and deserves further study, but for the moment we would concentrate on a subcategory **N** of **S** × **D**, where the **N**-objects are **S** × **D**-objects in which $S = S'$, i.e., of the form $((S, F), (S, D))$, and for convenience it will be denoted simply by (S, F, D).

Let us define the category **N** more explicitly:

6. Objects

An **N**-object is a triple (S, F, D), where (S, F) is an **S**-object and (S, D) a **D**-object.

We shall also consider S as a set on which different topologies can be defined. In particular, we shall consider the topological spaces (S, τ), where τ can be the f-topology for any $f \in F$ and the T-topology for any $T \in D$. Since we assume $0 \in F$ (0 denoting the zero function from S to $\mathbb{R}$) and $I_S \in D$ (the trivial dynamics), S is provided with both the indiscrete and the discrete topology [see V.4 and Example V.2(2)].

7. Morphisms

$$\phi \in \mathbf{N}((S_1, F_1, D_1), (S_2, F_2, D_2))$$
$$\text{if } \phi \in \mathbf{S}((S_1, F_1), (S_2, F_2)) \text{ and } \phi \in \mathbf{D}((S_1, D_1), (S_2, D_2)),$$

i.e., ϕ is a mapping of the sets $S_1 \to S_2$, $F_1 \to F_2$, and $D_1 \to D_2$ such that on (S, F) it satisfies the conditions of III.4 and on (S, D) it satisfies the conditions of IV.9. Note that a general $\mathbf{S} \times \mathbf{D}$-morphism (ϕ, ϕ') on $\mathbf{N}$-objects is not necessarily an $\mathbf{N}$-morphism, because for the latter ϕ and ϕ' have to "agree" on the state (phase) space S. So the inclusion functor $\mathbf{N} \to \mathbf{S} \times \mathbf{D}$ is faithful but not full.

We do not impose any relations for ϕ on (F, D). So, for example, $T \in D_1$ may be f-continuous for some $f \in F_1$, but $\phi T \in D_2$ is not required to be ϕf-continuous; and we can have T compatible with f but ϕT not compatible with ϕf, and so on.

8. Identity and Composition

The identity $\mathbf{N}$-morphism $1_{(S,F,D)}$ is clearly the amalgamation of $1_{(S,F)}$ and $1_{(S,D)}$, i.e., $1_{(S,F,D)}$ sends each $x \mapsto x \in S$, $f \mapsto f \in F$, and $T \mapsto T \in D$.

Composition of $\mathbf{N}$-morphisms is defined component-wise and is clearly associative with identity $1_{(S,F,D)}$.

If $\phi: (S_1, F_1, D_1) \to (S_2, F_2, D_2)$ is an $\mathbf{N}$-isomorphism, then (S_1, F_1) and (S_2, F_2) are $\mathbf{S}$-isomorphic, and (S_1, D_1) and (S_2, D_2) are $\mathbf{D}$-isomorphic (see Sections III.7 and IV.11). Note that even between $\mathbf{N}$-isomorphic systems, the continuity and compatibility properties of the dynamics and observables are not necessarily preserved. This is due to the fact that the observable are less well behaved and that $\sim$-equivalent observables, which are not necessarily topologically equivalent, are $\mathbf{S}$-isomorphic. This apparent shortcoming, contrariwise, turns out to be of great interest; some of these "bifurcation phenomena" will be discussed at the end of this section.

C. Constructions in N

9. Products and Coproducts

Since the phase (state) spaces of the $\mathbf{S}$-product and the $\mathbf{D}$-product of a family are the same (both being the **Ens**-product), we can construct the $\mathbf{N}$-product simply by "putting the two pieces together." Explicitly, let $\{(S_i, F_i, D_i) : i \in I\}$ be a family of $\mathbf{N}$-objects. Then the $\mathbf{N}$-product of this family is (S, F, D), where (S, F) is the $\mathbf{S}$-product of $\{(S_i, F_i)\}$ (when the $\mathbf{S}$-product exists) and (S, D) is the $\mathbf{D}$-product of $\{(S_i, D_i)\}$ (see Sections III.8 and IV.12).

Because of Theorem V.2(4), we see that continuous dynamics are preserved by $\mathbf{N}$-products. Also, it is easy to see that if the dynamics T_i is compatible with the equivalence relation R_i on S_i, then if the product dynamics $T = (T_i : i \in I)$ exists, it is compatible with the product equivalence relation R on S (where $(x_i : i \in I)R(y_i : i \in I)$ iff for each $i \in I, x_i R_i y_i$). In particular, if T_i is f_i-continuous for each $i \in I$, then $T = (T_i : i \in I) \in D$ is f-continuous ($f = (f_i : i \in I)$) on S, and vice versa.

Similarly, since the phase spaces of the **S**-coproduct and the **D**-coproduct are the same (both being the **Ens**-coproduct), the **N**-coproduct of the family $\{(S_i, F_i, D_i) : i \in I\}$ is (S, F, D), where (S, F) is the **S**-coproduct of $\{(S_i, F_i)\}$ and (S, D) is the **D**-coproduct of $\{(S_i, D_i)\}$ (see Sections III.11 and IV.13).

Continuous dynamics are preserved by **N**-coproducts [see Theorem V.2(5)]. Compatibility is also preserved with the coproduct equivalence relation on S defined by $(i, x)R(j, y)$ iff $i = j$ and $xR_i y$ in S_i.

10. Monomorphisms and Subobjects

It is clear that in a product category **A** × **B**, a morphism (ϕ, ϕ') is a mono (an equalizer, split mono, ...) if and only if ϕ is mono (an equalizer, split mono, ...) in **A** and ϕ' is mono (an equalizer, split mono, ...) in **B**. The corresponding statement, with obvious modifications, also holds for any subcategory of **A** × **B**. Further, we can have concepts like "partial monos" in **A** × **B**, where, for example, an **A**-mono in **A** × **B** is an **A** × **B**-morphism (ϕ, ϕ') such that ϕ is an **A**-mono and ϕ' is an arbitrary **B**-morphism.

Thus, an **N**-morphism is mono (an equalizer, split mono, ...) if it is such in both **S** and **D**. Now recall the hierarchies for monomorphisms (see Sections III.13 and IV.14).

in **S**: split mono $\Leftrightarrow$ equalizer $\Rightarrow$ mono $\Leftrightarrow$ injection,

in **D**: split mono $\Rightarrow$ equalizer $\Leftrightarrow$ mono $\Leftrightarrow$ injection.

The hierarchy for **N**-monomorphisms, taking the "intersection" of the two preceding statements, is then

split mono $\Rightarrow$ equalizer $\Rightarrow$ mono $\Leftrightarrow$ injection (on S, F, and D).

So there are three distinct types of monomorphisms in **N**; hence there are (at least) three possible definitions of an **N**-subobject. We shall choose to call an equalizer-subobject an **N**-subsystem because an equalizer preserves the relevant equivalence class structures on S (see Section III.17).

DEFINITION. Let $\phi \in \mathbf{N}((S_1, F_1, D_1), (S_2, F_2, D_2))$. Then (S_1, F_1, D_1) is an **S**- (respectively, **D**-, **N**-) *subsystem* of (S_2, F_2, D_2) if ϕ is an **S**- (respectively, **D**-, **N**-) equalizer [where ϕ is an **S**-equalizer if $\phi: (S_1, F_1) \to (S_2, F_2)$ is an **S**-equalizer and $\phi: (S_1, D_1) \to (S_2, D_2)$ is an arbitrary **D**-morphism, and so on].

All the different kinds of subsystems in the preceding definition share one decisive feature: the passage from a system to a subsystem places limitations on the interactive capabilities of the system. So an **S**-subsystem is limited in the measurements that can be performed on it (to obtain "observables") and hence is limited in the kinds of dynamics it can impose on other systems; a **D**-subsystem is limited in the kinds of dynamics that can be imposed on it and hence is limited in its capability as a meter; and an **N**-subsystem, which is *both* an **S**-subsystem and a **D**-subsystem, can be limited in both ways.

11. Epimorphisms and Quotient Objects

The hierarchies for epimorphisms are (see Sections III.14 and IV.16)

in **S**: split epi $\Rightarrow$ coequalizer $\Rightarrow$ epi $\Leftrightarrow$ surjection,

in **D**: split epi $\Rightarrow$ coequalizer $\Leftrightarrow$ epi $\Leftrightarrow$ surjection.

Therefore the hierarchy for **N**-epimorphisms is

split epi $\Rightarrow$ coequalizer $\Rightarrow$ epi $\Leftrightarrow$ surjection (on S, F, and D).

So there are three distinct types of epimorphisms in **N**, and hence many various definitions of **S**-, and **D**-, and **N**-quotient objects. (Incidentally, a *quotient object* is the dual concept to that of a subobject; a subobject is the domain of a monomorphism while a quotient object is the codomain of an epimorphism.)

We emphasize again that even in subobjects and quotient objects, the continuity properties of the dynamics are preserved only under special circumstances [see Lemma V.2(3) and Theorem V.2(6)].

12. Image Factorizations

N has both epi-equalizer and coequalizer-mono factorizations. The two factorizations are distinct in general, because of the distinct factorizations in **S** (see Sections III.16 and IV.16).

Let us consider first a general category **C** that has both epi-equalizer and coequalizer-mono factorizations. Let $\mathscr{E}$ = all **C**-epis and $\mathscr{M}$ = all **C**-equalizers. Let $f: A \to B$ in **C**, let $f = m_1 \circ e_1 : A \to X \to B$ be its epi-equalizer factorization, and let $f = m_2 \circ e_2 : A \to Y \to B$ be its coequalizer-mono factorization.

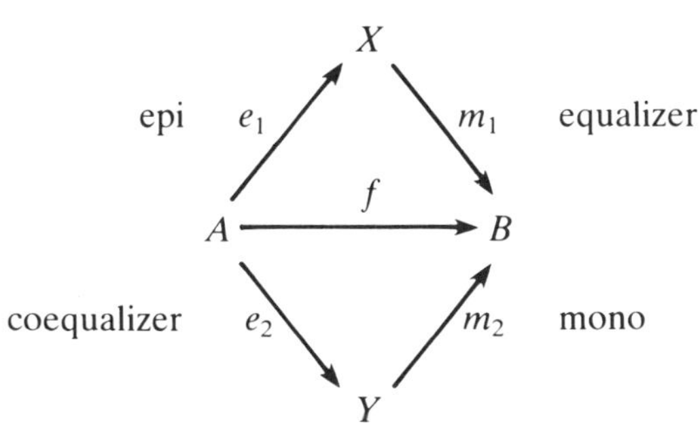

Then e_2 is a coequalizer hence an epi (general hierarchy) so $e_2 \in \mathscr{E}$; and $m_1 \in \mathscr{M}$. Thus by the diagonal fill-in lemma (see Section IV.15) there exists a unique $h: Y \to X$ with $e_1 = h \circ e_2$ and $m_2 = m_1 \circ h$. Further, since $e_1 = h \circ e_2$ is epi, so is h (see Section III.14); and since $m_2 = m_1 \circ h$ is mono, so is h (see Section III.13). In other words, there is a unique **C**-morphism h from the "coequalizer-mono image" Y to the "epi-equalizer image" X and h is both epi and mono (not necessarily an isomorphism, however).

Coming back to our category **N**, we see that for any $\phi:(S_1, F_1, D_1) \to (S_2, F_2, D_2)$, we have the commutative diagram

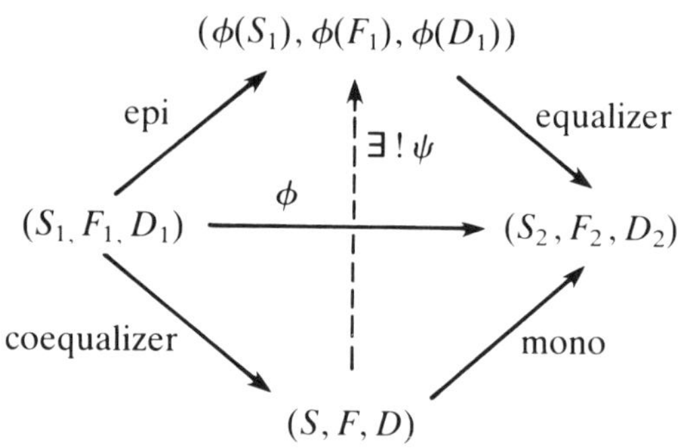

where ψ is both **N**-epi and **N**-mono but not necessarily an **N**-isomorphism. ψ is, however, a **D**-isomorphism from (S, D) to $(\phi(S_1), \phi(D_1))$ (see Section IV.16).

D. Bifurcations

13. Multiple Descriptions

The **N**-object (S, F, D) contains many different mathematical descriptions of the same system. There are many interesting questions about their connections one can ask: How do the descriptions of S obtained from one set of observables $\{f_1, f_2, \ldots, f_n\}$ compare with those from another set $\{g_1, g_2, \ldots, g_m\}$? How can one combine these descriptions to obtain a more comprehensive picture of S? To what extent does a knowledge that two states s_1 and s_2 appear "close" under the pseudometric on S induced by $\{f_1, f_2, \ldots, f_n\}$ imply that these same states appear close with respect to $\{g_1, g_2, \ldots, g_m\}$? How does a dynamics T appear when viewed through an observable f? And, vice versa, how does an observable appear after the passage of a dynamical process?

These questions are intimately related to the notions of *stability* and *bifurcation* (Thom, 1975). However, as we shall now see, questions of this type devolve once again to the notion of linkage (see Section III.9) and thence back to the essence of the modelling relation itself. Let us then take up the question of linkage of observables in more detail.

14. Linkage Reconsidered: Equations of State

Let us begin with the simplest case in which we have a set of states S with two observables $f, g \in \mathbb{R}^S$. We can represent this by a diagram

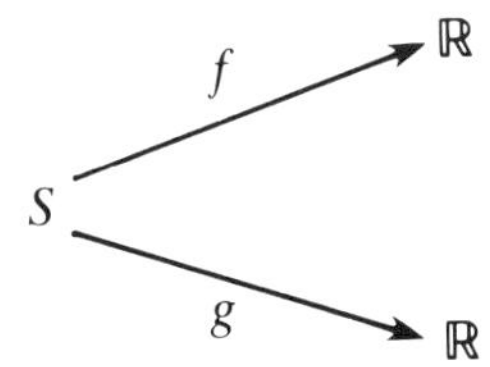

This is yet another type of the diagrams we discussed in Section VI.2. Here a single natural system is encoded into two formal systems, thereby establishing a correspondence between the two latter systems.

The linkage between two observables is, as we indicated in Section III.9, given by how much we learn about $g(s)$ when $f(s)$ is known, and conversely. In other words, linkage between f and g is manifested by the degree of correspondence between the two formal systems ($\mathbb{R}$ and $\mathbb{R}$) encoding S.

The simplest possible situation is when $R_f = R_g$. Then the value $f(s)$ of f at any $s \in S$ completely determines $g(s)$ and vice versa. This means there is actually an (**Ens**-) isomorphism Ψ between $f(s)$ and $g(s)$ [i.e., $(S, \{f\})$ is isomorphic to $(S, \{g\})$ in **S**] and we have for every $s \in S$

$$\begin{aligned} \Psi(f(s)) &= g(s) \\ f(s) &= \Psi^{-1}(g(s)). \end{aligned} \tag{1}$$

Diagrammatically,

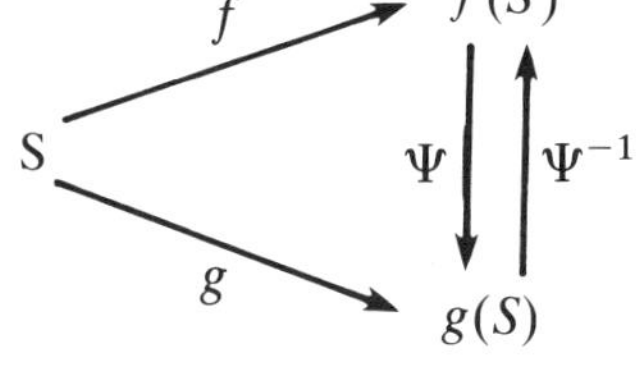

Alternatively, when $R_f = R_g$, we have $R_{fg} = R_f = R_g$ and the image of S/R_{fg} under the embedding $(s)_{fg} \mapsto ((s)_f, (s)_g)$ is simply a "graph" in $S/R_f \times S/R_g$ $(= f(S) \times g(S))$, whose equation is given by (1).

The next situation we can consider is when R_f refines R_g (i.e., when g is totally linked to f). In this case each class in S/R_g is a union of classes in S/R_f, and knowing the value $f(s)$ completely determines $g(s)$, but not vice versa in general. Here $R_{fg} = R_f$ and there is a noninvertible (many-to-one) mapping $\Psi: f(S) \to g(S)$. In other words, the g-encoding may be *reduced* to the f-encoding, but not conversely.

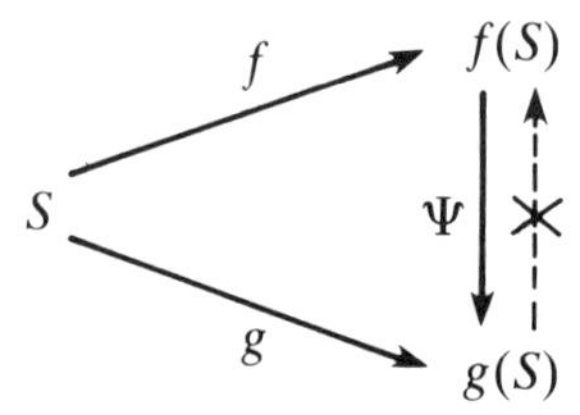

Geometrically, this means that the image of S/R_{fg} in $f(S) \times g(S)$ is a graph satisfying a relation, for all $s \in S$

$$\Phi(f(s), g(s)) = 0, \tag{2}$$

in which $f(s)$ plays the role of an independent variable; i.e., Eq. (2) can be solved for g as a single-valued function of f (e.g., when Φ is differentiable, $\partial\Phi/\partial f \neq 0$) but not conversely (for example, $\partial\Phi/\partial g = 0$).

More generally, if g is partially linked to f, i.e., when the value $f(s)$ of f partially constrains (imposes "selection rules" on) the possible values of $g(s)$, then the image of S/R_{fg} in $f(S) \times g(S)$ still satisfies a (nontrivial) relation of the form (2), except now we can solve for neither f nor g as a single-valued function of the other. Here there is no mapping between encodings in either direction and so neither of them can be regarded as contained in the other.

Finally, when f and g are unlinked, the map $S/R_{fg} \to f(S) \times g(S)$ is onto, and there is no meaningful relation of the form (2).

Thus we see that a linkage between two observables f and g can be expressed as a relation of the form (2), which characterizes some subset of $f(S) \times g(S)$. Such a relation is called an *equation of state* for the system. It is a relation between encodings of a natural system, expressing the degree to which the encodings are linked. An equation of state is not an observable but rather represents the encoding of *system laws*; i.e., it is a rule of inference corresponding to a system behaviour.

The preceding discussion can easily be generalized to two families of observables $F = \{f_1, f_2, \ldots, f_n\}$ and $G = \{g_1, g_2, \ldots, g_m\}$ on S. Any linkages between the two families (and in fact within the families) can be expressed by an equation of state of the form

$$\Phi(f_1(s), \ldots, f_n(s), g_1(s), \ldots, g_m(s)) = 0. \tag{3}$$

The character of Φ (e.g., the vanishing of some of its derivatives) will express the manner in which the observables are linked. Categorically, this idea is

expressed in the product diagram

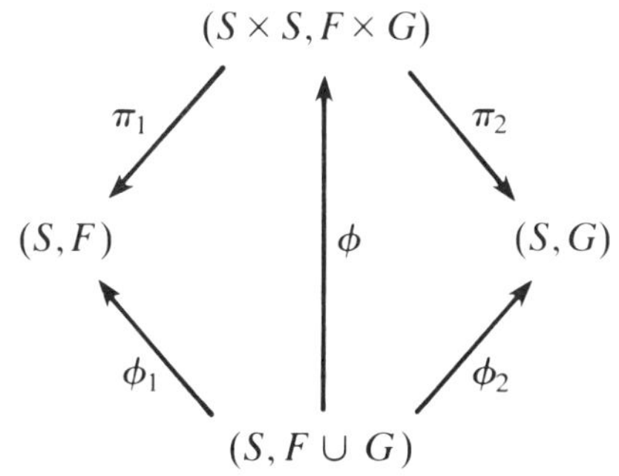

and the linkage is reflected in the nature of the S-morphism ϕ.

15. Stable Points

In the preceding section we considered those relationships among observables $f, g, \ldots$ based on the concept of linkage, which depends only on the equivalence classes of the relations $R_f, R_g, \ldots$. We considered the manner in which a "state transition" $s \to s'$ that was undetectable by f (i.e., $f(s) = f(s')$) could be detected by g (i.e., $g(s) \neq g(s')$).

Next, we shall try to answer this question: If s is "close" to s' under (the pseudometric f-topology induced by) f, when will s also be close to s' under g? Thus we are considering the extent to which a state transition $s \to s'$ that is "almost" undetectable by f (i.e., $|f(s) - f(s')|$ is small) is likewise almost undetectable by g (i.e., $|g(s) - g(s')|$ is small).

DEFINITION. Let f, g be observables on S and d_f, d_g be the corresponding induced pseudometrics (see Section V.4). The state $s \in S$ is a *stable point of g with respect to f* if for every $\varepsilon > 0$ there exists a $\delta > 0$ such that for $s' \in S$, $d_f(s, s') = |f(s) - f(s')| < \delta$ implies $d_g(s, s') = |g(s) - g(s')| < \varepsilon$.

The definition is equivalent to each of the following:

(i) the identity map of the set S from (S, d_f) to (S, d_g) is continuous at s;

(ii) the f-open neighbourhood system of s refines the g-open neighbourhood system of s;

(iii) (roughly) every state f-close around s is also g-close around s.

The following is clear:

LEMMA. (a) If $s \in S$ is a stable point of g with respect to f then $(s)_f \subset (s)_g$.

(b) If $s \in S$ is a stable point of g with respect to f and $s' \in (s)_f$, then s' is also a stable point of g with respect to f. $\square$

THEOREM. The set of stable points of g with respect to f is an f-open subset of S. □

16. Bifurcation Set

DEFINITION. The complement of the set of stable points of g with respect to f is the *bifurcation set* of g with respect to f.

Intuitively, near a bifurcation point of g with respect to f, the proximity of two states s and s' of S as viewed by the observable f does not imply their proximity as viewed by g. In other words, the bifurcation set is the set of states at which the g-description does not agree with the f-description in their metrical aspects, i.e., the two descriptions convey essentially different information.

It follows from Theorem VI.15 that the bifurcation set of g with respect to f is an f-closed subset of S. Also, it is clear that the bifurcation set of g with respect to f is empty if and only if the f-topology is finer than the g-topology on S, in which case R_f refines R_g.

17. Equivalence of Observables

In the previous discussion, we can interchange the roles of f and g and obtain the opposite concept of stable and bifurcation points of f with respect to g. These are in general quite different from those obtained from g with respect to f. Thus, given a pair of observables, we obtain two distinct notions of stability and bifurcation, depending on which description is chosen as the reference.

Let us consider the case when f and g are two observables on S such that the bifurcation sets of f with respect to g and of g with respect to f are both empty. Then $1_S\colon (S, d_f) \to (S, d_g)$ is a homeomorphism and d_f and d_g are equivalent pseudometrics, i.e., f and g induce the same topology on S. Under these circumstances it is appropriate to say that f and g are *topologically equivalent* (as opposed to *algebraically equivalent* $f \sim g$, when $R_f = R_g$). Note that by Lemma VI.15, topological equivalence implies algebraic equivalence; but not conversely. Stated another way, we have the following theorem.

THEOREM. If two observables induce the same topology on the set of states, then they are totally linked to each other. □

18. Bifurcation and Continuity

Let us suppose that the two observables f, g on S are related by an equation of state of the form

$$\Phi(f(s), g(s)) = 0 \qquad (1)$$

for all $s \in S$. We shall consider how the concepts of stability and bifurcation are reflected in some (if any) properties of the function $\Phi: f(S) \times g(S) \subset \mathbb{R}^2 \to \mathbb{R}$.

Let $s \in S$ and let U be a neighbourhood of $f(s)$ in $\mathbb{R}$. If $g \circ f^{-1}(U)$ is a small neighbourhood of $g(s)$ whenever U is small, then s is a stable point of g with respect to f. The relation of $g \circ f^{-1}(U)$ to U is clearly another way of expressing the equation of state (1) (around the point $(f(s), g(s))$). It is also clear that s being a stable point of g with respect to f is equivalent to the continuity of the second argument of Φ with respect to the first at $(f(s), g(s))$. Dually, a stable point of f with respect to g occurs at a point of continuity of the first argument of Φ with respect to the second. In other words, bifurcation depends on, and is a consequence of, the discontinuity of the linkage relation between observables, when the properties of distinct encodings become logically independent.

Again, the preceding can be generalized to linkages among more than two observables.

19. Incompatibility

We saw in Section IV.18 that when a dynamics T on a set S is compatible with an equivalence relation R on S, T induces a quotient dynamics on the set of reduced states S/R. In this section we shall investigate what happens when a dynamics T is *not* compatible with an equivalence relation R, in the special case when $R = R_F$ where F is a family of observables on S.

If T is not compatible with R_F, then there are states $s, s' \in S$ for which sR_Fs' but which for some $t \in \mathbb{R}$ T_ts and T_ts' are not R_F-related. That is, T splits equivalence classes of R_F. Putting it another way, the two indistinguishable states s and s' (under F) have now *differentiated* through the action of the dynamics T, and that this differentiation is visible using the observables in F. (The term "differentiation" is used deliberately here to suggest the connection of this with biological differentiation. See Section VII.5.)

From the viewpoint of an observer equipped with meters for observables in F, the states s and s' appeared to be the same. But through the course of the dynamics T the observer detects two different states T_ts and T_ts'. It would appear that the same initial state under the same conditions has given rise to two distinct states, a contradiction to causality. The problem here is of course that one usually assumes that one has a complete set of observables F for the description of S (i.e., $S/R_F = S$, or R_F is the equality relation). The standard way out is to pull in statistics and to observe many copies of $(s)_{R_F}$ under the passage of T. The relative frequencies of the resulting states $(T_ts')_{R_F}$ [where $s' \in (s)_{R_F}$] are then associated with *transition probabilities* from s to T_ts'. In other words, the incompatibility of T with R is usually interpreted in stochastic terms.

One can (rather boldly) make the suggestion that all processes occurring in nature are deterministic, and the apparent stochasticity is simply a consequence of employing an incomplete description as if it were complete. So one does not need statistical tools *if* one has a complete description of the system. But that is a rather big *if* because we are limited in our means of observation, measurement, and understanding, and to obtain a complete description of every natural system is really to find the philosopher's stone! Thus statistics plays a role in science as a matter of necessity. Further discussions of the interplay between causality and chance can be found in Bohm (1957) and in Belinfante (1973) on the theory of "hidden variables."

20. *Stability and Bifurcation, Again*

We just considered the situation when a change of state $s \to s'$ that is undetectable by an observable f becomes detectable by the observable $g = f \circ T_t$. This of course is again intimately related to the concept of linkage. After all, given an observable f and a dynamics T on S, for each $t \in \mathbb{R}$ $f \circ T_t$ is an observable on S. To say that T is compatible with R_f (i.e., f) simply means that R_f refines each $R_{f \circ T_t}$ or that each $f \circ T_t$ is totally linked to f.

With this in mind, the next natural question to ask is: If s is close to s' under f, when will s also be close to s' under f after the action of a dynamics T? This problem can then clearly be studied by reducing to consider stable and bifurcation points of $f \circ T_t$ with respect to f. And the results from the previous sections can be appropriately modified and used.

Since the compatibility of T with f is equivalent to the commutativity of the following diagram (see the remark of Section IV.18).

$$\begin{array}{ccc} \text{dom } T & \xrightarrow{\chi \times 1_{\mathbb{R}}} & \text{dom } T' \\ {\scriptstyle T}\downarrow & & \downarrow{\scriptstyle T'} \\ S & \xrightarrow[\chi]{} & S/R_f \end{array}$$

the study of stability and bifurcation can be formulated in the category **N** as the "approximate commutativity" of this diagram.

Finally, there is an interesting possibility that a state $s \in S$ can be a stable point of $f \circ T_t$ with respect to f for all t less than a critical time t_c, and then for $t > t_c$, s becomes a bifurcation point of $f \circ T_t$ with respect to f. In other words, at $t = t_c$ we have a *catastrophe* (in the sense of Thom, 1975). Alternatively we can consider $t \in \mathbb{R}$ as an order parameter and at $t = t_c$ we have a change of scheme from an old structure to a new structure through an instability. This area is a further topic of investigation and we shall not deal with it here. A good set of reference is the Springer Series in Synergetics, especially the introductory Volume 1 (Haken, 1978) and Volume 4 (Güttinger and Eikemeier, eds., 1979), *Structural Stability in Physics.*

VII. BIOLOGICAL IMPLICATIONS

> Any particular or isolated biological phenomenon or group of phenomena admits of necessity an explanation in terms of a mathematical model.
>
> Nicolas Rashevsky

We come now to the final subject of our study. We want to show how the mathematical formalism of categorical system theory we developed may be applied in the analysis of various biological problems.

A. Development and Senescence

1. The Nature of Biology

Biology is the science of life and of the living, and of the multitude of interactions among living systems. The term *life* encompasses an immense set of phenomena; thus biology seems to sit at the rendezvous of all other sciences. The physicist Yang Chen-Ning once said that physics and mathematics are like two leaves sharing the same petiole, that they do not overlap but rather have the same blood in their veins. We can generalize the idea and say that science is like a compound leaf; each leaflet is one branch of science sharing the same rachis, which represents biology, and that biology is a unique science in the sense that this compound leaf is itself a biological object, a living system.

What, then, is a living system and what does it do? Many scientists coming from diverse backgrounds, when engaged in the search for general principles to integrate our understanding of the phenomena of life, have placed major emphasis on the notion of life as viewed from the standpoint of their own specialties. This is a classic example of a reductionist's view of the world. However, although the various theories of living systems differ greatly in their concepts and definitions of basic terms, they seem to share a common goal: to organize the findings (observations!) in some or all of biology into a single conceptual structure.

2. Developmental Biology

The development, the progressive production of the phenotypic characteristics, of organisms is a mysterious process (Berrill and Karp, 1976; Trinkaus,

1969). How can a single cell, the zygote, change itself into an adult containing millions of cells organized in a complex way? The fact that all the information necessary to produce the adult is contained in the very first cell and that the environment simply provides energy and materials increases the difficulty of the problem. The zygote of a given species always becomes an adult of that species and never that of another, whatever its environment. Growth mainly occurs through a continuous process of cell division, but how does differentiation that occurs during growth come about? Since all the cells contain the same genetic instructions, how do they come to differ from one another structurally and functionally and develop into elaborate spatial patterns? How does a developing embryo repair damages that may occur accidentally? Surely precise mechanisms for control and adaptation are involved—but *how*? Perhaps the solution to these complex problems of developmental biology lies not in modelling the phenomena but rather in abstract formal treatments like automata theory (see, for example, Arbib, 1972) and our categorical system theory.

Developmental biology has become an immense field whose boundaries are difficult to define (Thompson, 1942). The focus has been and remains on the embryo—on the gradual emergence of form and structure. The essence of embryonic development is *change*—transition from one state to another (dynamics!). Embryos are a "fleeting state," and development is an expression of the irreversible flow of biological events along the axis of time.

The transformation of an organism or its component cells from one state to another can also be identified in a variety of biological processes not specifically related to embryonic development. Of prime interest among them are aging, regeneration, and malignancy, each of which is characterized by a shift, either gradually or suddenly (catastrophically?), in cellular activities. These processes also fall into the general domain of development biology.

The article by Crick (1977) is an enlightening one about our ignorance—what we do *not* know—of development biology.

3. *Aging*

The outward signs (i.e., observables) of aging are obvious; yet the underlying bases for the deteriorative processes that occur in all living organisms are very poorly understood. In fact, the very definition of the term *aging* is far from clear; it may well be that *aging* is a *primitive* of a logical theory and hence is left as an undefined term. Primitives receive no definitions except those given to them implicitly by their presence in the general axioms of the theory. In our study, the terms *system*, *state*, and *observable* are primitives (see Section III.1). But can we define what aging is in our framework of categorical system theory?

There is no single observable that can be called aging, although many basic observations have been made of age-related changes [i.e., those changes correlated with increasing (chronological) age]. In fact, almost every observable monotone function of chronological age has been used as a measure of biological age. [Note this statement itself assumes that somehow biological age is comparable to chronological age, i.e., time. So just as in dynamical systems (see Section V.9), we are assuming the existence of a "scaling factor" between intrinsic age and extrinsic time.] For example, increasing functions from blood pressure to amount of lipofucsin in cells, and decreasing functions like organ activity, nerve conduction velocity, muscle power, etc., have all been used as age indicators. One of the most important tasks of aging research is to determine which of these changes are primary events that could be directly responsible for survival and which are either secondary manifestations or independent processes not responsible for causing age-related death.

Some investigators believe that aging, whether caused by intrinsic or extrinsic factors, is a general property of all normal cells. Others have proposed that a limited population of cells controls the course of aging throughout the (multicellular) organism. Above all, perhaps it should be emphasized that *aging and growth are inseparable.* Growth as such may be analyzed in terms of cell multiplication, cell enlargement, cell replacement, and other features such as accumulation of extracellular substance. Yet the phenomenon itself, including factors that determine both the rates of growth and the limits of growth, is not understood. Aging is essentially an extension of the growth process into negative values; i.e., the process is seen as increase as long as cell multiplication exceeds cell death and as decrease when cell death exceeds cell multiplication. The process is continuous throughout life, beginning in early development. When the underlying growth-controlling properties become better understood, so will the phenomena of aging and death.

Let us now return to our formal treatment.

4. *A Partial Order in* **N**

DEFINITION. Let (S_1, F_1, D_1) and (S_2, F_2, D_2) be **N**-objects. Then $(S_1, F_1, D_1) \leq (S_2, F_2, D_2)$ if

(i) there is an **N**-mono $\phi: (S_1, F_1, D_1) \rightarrow (S_2, F_2, D_2)$;
(ii) S_1, F_1, and D_1 are finite sets.

Note first that if $(S_1, F_1, D_1) \leq (S_2, F_2, D_2)$, then $(S_1, F_1) \leq (S_2, F_2)$ in **S** (see Section III.18). Also, since S_1 is a finite set, for each $f \in F_1$ $f(S_1)$ is a finite subset of $\mathbb{R}$; hence any representative of the class $(f)_\sim$ induces the same f-topology on S_1, i.e., there is a unique $(f)_\sim$-topology on S_1 (see Section V.4).

It is clear that $\leq$ on the finite **N**-objects [i.e., on (S, F, D), where S, F, and D are finite sets] is reflexive and that $\leq$ is transitive. It is also clear, using a similar argument as in Section III.18, that if $(S_1, F_1, D_1) \leq (S_2, F_2, D_2)$ and $(S_2, F_2, D_2) \leq (S_1, F_1, D_1)$ then the **N**-monos involved are in fact **N**-isomorphisms. So $\leq$ is antisymmetric (up to isomorphism, as usual). Therefore $\leq$ is a partial order on (the isomorphism classes of) the finite **N**-objects.

Let $(S_1, F_1, D_1) \leq (S_2, F_2, D_2)$. Since the phase space S_1 is finite, each $T \in D_1$ has dom $T = S_1 \times \mathbb{R}$ (see Corollary 2 in Section IV.6); i.e., for every $x \in S_1$ $a_x[T] = -\infty$ and $b_x[T] = +\infty$. But then from $a_{\phi x}[\phi T] \leq a_x[T] < 0 < b_x[T] \leq b_{\phi x}[\phi T]$, we have $a_{\phi x}[\phi T] = -\infty$ and $b_{\phi x}[\phi T] = +\infty$ as well, i.e., dom $\phi T = S_2 \times \mathbb{R}$. So here we are simply dealing with the usual concept of dynamics. Since dom $T = S_1 \times \mathbb{R} \rightarrowtail S_2 \times \mathbb{R} =$ dom ϕT, T can be considered as the relative dynamics on S_1 induced by the dynamics ϕT on S_2; alternatively, ϕT can be considered as an *extension* of the dynamics T.

5. *Growth and Aging as a Partial Order*

What does this partial order $\leq$ in **N** have to do with our present topic on development and senescence? Since $(S_1, F_1, D_1) \leq (S_2, F_2, D_2)$ in **N** implies in particular that $(S_1, F_1) \leq (S_2, F_2)$ in **S**, we can include the discussions from Section III.18. Let us see what we have.

With an **N**-mono $\phi: (S_1, F_1, D_1) \leq (S_2, F_2, D_2)$, the possibility that ϕ is not onto models *growth*. If $S_2 \sim \phi(S_1)$ is not empty, then more elements (states) have appeared in the second system, an increase in *size*. If $F_2 \sim \phi(F_1)$ is nonempty, then there are more observables in the second system, an increase in *complexity*. If $D_2 \sim \phi(D_1)$ is nonempty, then more dynamics can be imposed on the second system, an increase in *interactive ability*. Collectively, the appearance of these new modes of structure, organization, and behaviour falls into the description of the sometimes puzzling biodynamical phenomenon termed *emergence*.

When $g \in F_2 \sim \phi(F_1)$ and $s, s' \in S_1$ are such that $sR_{F_1}s'$ but $g(\phi s) \neq g(\phi s')$, indistinguishable states in S_1 are now separated in S_2 because of an increase in complexity, an alternate description. When $T \in D_1$ is compatible with R_{F_1} but ϕT is not compatible with some $g \in F_2$ (see Section VI.19), we have the case that indistinguishable states in S_1 become separated in S_2 because of an interaction through an additional dynamics. Both of these cases indicate the presence of *differentiation* going from (S_1, F_1, D_1) to (S_2, F_2, D_2).

Suppose now $s, s' \in S_1$ and $f \in F_1$ are such that $f(s) \neq f(s')$ yet $(\phi f)(\phi s) = (\phi f)(\phi s')$, then distinct states become "the same." Also, when a dynamics is not compatible with an equivalence relation, different equivalence classes may appear to "fuse" together in the course of the dynamical process. These serve as models for biological *integration* when interpreted "positively," and for *decay* (loss of information) when interpreted "negatively."

We mentioned (Section V.9) that the $1_{\mathbb{R}}$ in the map $\phi \times 1_{\mathbb{R}}: \text{dom}\, T \to \text{dom}\, \phi T$ is a simplified version of the time scaling in which we have made the identification $t_1 \leftrightarrow t_2$. Suppose instead of $1_{\mathbb{R}}$ we use a monotone increasing function $a: t_1 \mapsto t_2$ from $\mathbb{R}$ to $\mathbb{R}$, then the map $T \mapsto \phi T$ with the commutativity condition $\phi T(\phi x, a(t)) = \phi(T(x, t))$ would have incorporated into it a "speeding up" or "slowing down" of the dynamics depending on whether $\dot{a} > 1$ or $\dot{a} < 1$. This again models growth and decay.

When a change of continuity properties occurs going from (S_1, F_1, D_1) to (S_2, F_2, D_2)—for example, when T is f-continuous ($T \in D_1$, $f \in F_1$) but ϕT is not ϕf-continuous, or when the linkage between ϕf and ϕg is different from that between f and g ($f, g \in F_1$)—it could be interpreted as a change of biological structures and functions. This kind of apparently discontinuous change in biological systems again falls into the area of "emergence." The generation of emergent novelties is highly characteristic of biological systems, and in our formalism of categorical system theory it is a natural consequence of the definition of $\leq$.

With all the preceding in mind, we can now make a formal definition of aging and growth:

DEFINITION. Let (S_1, F_1, D_1) and (S_2, F_2, D_2) be finite natural systems. Then (S_1, F_1, D_1) is *younger* than (S_2, F_2, D_2) (and the latter is *older* than the former) if $(S_1, F_1, D_1) \leq (S_2, F_2, D_2)$; i.e., *aging* is defined as the partial order $\leq$ on the finite natural systems.

6. *The Directionality of Aging*

There is probably no subject that so deeply interests human beings as that of the duration of human life. More than most sciences, gerontology is haunted by primordial myths and fears. This concern is extremely ancient. Somehow people always seem to believe that there are magical ways of reversing, or at least postponing (preferably forever), the process of aging. There are numerous examples in history of searches for these magical ways: Xu Fu was sent by the Chinese emperor in the third century B.C. to look for the Fairy Islands (incidentally, this is supposed to be the origin of Japan); Faust sold his soul to the Devil in exchange for a promise of immortality, knowledge, and power; and the Spanish conquistador Ponce de León searched for the Fountain of Youth in the sixteenth century (but was shot down by native arrows in Florida).

The goals of modern aging research are considerably more modest. We are still at the stage where we are trying to *understand* aging. In fact, the more we know about aging, the more irreversible a process it seems to be. That is why ideas from irreversible thermodynamics and dissipative systems are used to model aging (see, for example, Richardson, 1980). The irreversibility of aging is captured in Definition VII.5. If $(S_1, F_1, D_1) \leq (S_2, F_2, D_2)$ then of course in general we do not have $(S_2, F_2, D_2) \leq (S_1, F_1, D_1)$. Indeed, if the latter holds we would have $(S_1, F_1, D_1) \cong (S_2, F_2, D_2)$ in **N**, in which case the two systems are of the "same age." So the partial order $\leq$ gives a (uni)directionality of aging and growth.

Note, however, that although in general an **N**-mono does not split (see Section VI.10)—this is what we need for $\phi: (S_1, F_1, D_1) \rightarrowtail (S_2, F_2, D_2)$ to be reversible—it is possible that there is a subsystem (S, F, D) of $(\phi(S_1), \phi(F_1), \phi(D_1)) \subset (S_2, F_2, D_2)$ on which ϕ^{-1} exists and is an **N**-mono. In other words, aging results from properties and relations of whole systems, and it does not forbid the possibility that one or more of the component subsystems give opposite contributions. Aging is a *collective* (*cooperative*) *phenomenon* of many processes, some of which may appear to defy aging.

B. The Concept of the Organism

7. *Organism*

When one thinks of collective phenomena in which the discrete constitutive individuals are modified in their behaviour through interactions among one another to fit into the patterns of a larger collective set, and the whole is more than and different from a simple addition of its parts, living organisms would seem to be the ideal example. Yet the concept of the organism has resisted all attempts of definition. Partly it is because organisms are complex, which means that they admit many different kinds of descriptions. The characteristics of any of the many possible descriptions made of an organism provide a possible way of defining the organism. But many of these descriptions look contradictory and hence add to the intrinsic difficulty of the task. The various academic fields that study organisms approach their tasks quite differently. None has a complete picture, and communication among these disciplines is often poor. Gross anatomy concerns structure; general physiology primarily deals with matter–energy processing; and endocrinology, neurophysiology, genetics, and psychology chiefly deal with information processing. On the other hand, since these different descriptions represent different kinds of abstractions from real organisms, each can tell us something that the others cannot. Perhaps the most important problem lies in trying to combine these divergent views of the organism into a more comprehensive picture. (Collective description of a collective phenomenon?)

The conventional (reductionistic) view of the organism is that of a physical-chemical system whose behaviour is to be deduced from its structures according to the laws of physics and chemistry. Life, according to this view, is a potentially self-perpetuating open system of linked organic reactions, catalyzed stepwise and almost isothermally by complex and specific organic catalysts that are themselves produced by the system. Another view of the organism is from that of organic behaviour, in the context of a relational biology, as conceived originally by Nicolas Rashevsky (see Rashevsky, 1960). On the integrative aspects of behaviour, it was Rashevsky's idea that the organisms are recognized as such because we can observe homologies in their behaviours, regardless of the physical structures through which these observations are made. Thus all organisms manifest the same set of basic and ubiquitous biological functions, and through this manifestation organisms could be mapped on one another in such a way as to preserve these basic relations. This idea led to the formulation of Rashevsky's principle of biotopological mappings and Robert Rosen's (categorical) (M, R)-systems (Rosen, 1972). On the adaptive and predictive character of organic behaviour, one is led to the classical (optimal) control theory (Rosen, 1980), and Rosen's theory of anticipatory systems (Rosen, 1985).

The following is a description of the developmental processes of an organism from a categorical standpoint.

Definition. Let $\mathbf{L}$ consist of

(1) a collection $\{L_i = (S_i, F_i, D_i) : i \in I = [0, 1]\}$ of $\mathbf{N}$-objects such that for $i \leq j$ in I, $L_i \leq L_j$ in $\mathbf{N}$. (So in particular the L_i's are 'finite' $\mathbf{N}$-objects.)

(2) for each pair $i, j \in I$ with $i \leq j$, a hom set $\mathbf{L}(L_i, L_j)$ containing a *single* $\mathbf{N}$-mono $\phi_{ji}: L_i \to L_j$, such that for $i \leq j \leq k$ in I, $\phi_{kj} \circ \phi_{ji} = \phi_{ki}$

$$L_i \xrightarrow{\phi_{ji}} L_j \xrightarrow{\phi_{kj}} L_k, \qquad \phi_{ki}: L_i \to L_k$$

and if $i > j$ in I then there is no $\mathbf{N}$-morphism in $\mathbf{L}$ from L_i to L_j; i.e., $L(L_i, L_j) = \varnothing$ for $i > j$.

It follows from (2) that for each $i \in I$, $\{\phi_{ii} = 1_{L_i}\} = \mathbf{L}(L_i, L_i)$. Thus $\mathbf{L}$ is a subcategory of $\mathbf{N}$ and the set of $\mathbf{L}$-objects is totally ordered by $\leq$. In fact, one easily sees that $\mathbf{L}$ is isomorphic (in **Cat**) to the totally ordered set $(I, \leq)$ considered as a category [see Example II.4(6)].

Definition. An *organism* is a natural system that is (a realization of) $\mathbf{L}$ for an appropriately chosen family $\{L_i\}$ of $\mathbf{N}$-objects and an appropriately chosen collection $\{\phi_{ji}\}$ of $\mathbf{N}$-monos satisfying the definition of $\mathbf{L}$.

Considering the results we have from the discussions on aging, it is quite reasonable to make the preceding definition of an organism from the standpoint of categorical system theory. The totally ordered set $(I, \leq)$ is an index of *age* and the order $\leq$ on $\mathbf{L}$ is the process of aging. The instant $i = 0$ can be considered as the moment of *conception* of an organism (when *life* begins) and the instant $i = 1$ is *death*. The developmental processes of the organism are reflected in the systems (S_i, F_i, D_i), the morphisms ϕ_{ji}, and in particular in the evolution of the systems as the index $i \in I$ increases from 0 to 1.

It is an appropriate place here to mention a new subject of study started by Rosen (1985), that of *anticipatory systems*. The basis for this theory is the recognition that most (if not all) of the biological behaviour is of an anticipatory rather than a reactive nature. Let us see if our definition of an organism has incorporated into it this anticipatory character. From the total order on the set of **L**-objects we can conclude that certain things *cannot* happen as the organism develops (i.e., as i increases). For example, it cannot happen that $f(s) = f(s')$ but $(\phi_{ji} f)(\phi_{ji} s) \neq (\phi_{ji} f)(\phi_{ji} s')$, it cannot happen that $T \neq T'$ but $\phi_{ji} T = \phi_{ji} T'$, and so on. So this list of impossible happenings can be considered as a prediction of things to come (or rather, *not* to come). Also, the linkage imposed on the states by the dynamics coupled with causality determines how a present state can be regarded as a model of a future state. There are many qualities like these one can list. So perhaps a formal study of the "category of anticipatory systems" is a perspective for the future. (Note this last statement is itself of an anticipatory character!)

8. *Organismic Observables*

An observable represents a capacity for interaction manifested by the ability to move a meter. Interactions and indicators are abundant at the level of the organism, especially for human organisms. Let us list a few—the trend is that the more "advanced" the organism, the more observables (where applicable) there are.

Organismic observables (Comorosan, 1976) include measurements of external appearance such as size, posture, colour, and deformity, and external behaviour such as speech and motor functions. They also include such physiological conditions as body temperature, pulse, blood pressure, rate of respiration, brain electrical activity, excretion, sensory functions, and reflexes. There are also very much meter-dependent observables such as appearance under x-ray, fluoroscope, and microscope. (This gives an example of how we continually learn about natural systems as our means of observation and understanding grow.) For highly advanced organisms we can include psychological indicators such as intelligence tests, performance tests, etc. Social organisms can be classified according to social status, roles, and so on. At the present stage of the development of biology, however, the most reliable and precise observables (i.e., those that give us the most complete description of the organism) are probably the biochemical ones. Normal limits and extreme ranges of hundreds of biochemical variables have been determined, methods of measurement specified, instruments perfected, and the meaning of changes in values established. Molecular biology is undoubtedly at the present time the central dogma of biology. Nevertheless, let us not forget what Daniel Bernoulli once said: There is no philosophy which is not founded upon knowledge of the phenomena, but to get any profit from the knowledge it is absolutely necessary to be a mathematician. Mathematical biology shall come of age as a matter of necessity.

C. Organismic Sets and Living Systems

9. Organismic Sets

Organismic sets were built by Nicolas Rashevsky as a representation of biological organisms and societies on a relational basis (Rashevsky, 1972), and a wide range of biological and social phenomena were explained within this framework.

The idea was first started by the observation (!) of the remarkable relational similarities among physics, biology, and sociology. The phenomena are properties of collections of things that are capable of performing certain activities that result in certain products (a cause-and-effect phenomenology!). This led to the suggestion of the existence of a conceptual superstructure of which physics, biology, and sociology are three parallel branches, each partially isomorphic to the other two. (Note the resemblance of the idea of Section VII.1 to this.) This conceptual superstructure is an *organismic set*.

Even before we formally define what an organismic set is, we can see that the setting is perfect for a category theory to be postulated. After all, we are looking at a class of mathematical objects with the same structure. Let us first take a brief digression into considering the idea of *structure* in mathematics in terms of categories.

10. Structured Categories

DEFINITION. Let $\mathbf{C}$ be a fixed category (usually **Ens**). A *category of* $\mathbf{C}$-*objects with structure*, $\mathbf{K}$, is given by the following:

(1) $\mathbf{K}$ assigns to each $\mathbf{C}$-object X a class $\mathbf{K}(X)$ of $\mathbf{K}$-*structures on* X. A $\mathbf{K}$-*structure* is a pair (X, σ) with $\sigma \in \mathbf{K}(X)$.

(2) To each ordered pair of $\mathbf{K}$-structures $((X, \sigma), (Y, \tau))$. $\mathbf{K}$ assigns a subset $\mathbf{K}(\sigma, \tau)$ of $\mathbf{C}(X, Y)$ of $\mathbf{K}$-*admissible* $\mathbf{C}$-*morphisms from* (X, σ) *to* (Y, τ). For $f \in \mathbf{K}(\sigma, \tau)$ we write $f: (X, \sigma) \to (Y, \tau)$.

They satisfy the following:

(i) *Axiom of composition:* If $f: (X, \sigma) \to (Y, \tau)$ and $g: (Y, \tau) \to (Z, \upsilon)$ then $g \circ f: (X, \sigma) \to (Z, \upsilon)$.

(ii) *Structure is abstract*: If $f: X \to Y$ is a **C**-isomorphism, then for all $\tau \in \mathbf{K}(Y)$ there exists a unique $\sigma \in \mathbf{K}(X)$ such that $f:(X,\sigma) \to (Y,\tau)$ and f^{-1}: $(Y,\tau) \to (X,\sigma)$.

REMARKS. It follows from the definition that $1_X:(X,\sigma) \to (X,\sigma)$ for all $\sigma \in \mathbf{K}(X)$. Thus there is a category, denoted by **K**, with objects all **K**-structures and morphisms all **K**-admissible **C**-morphisms. There is also obviously the *underlying* **C**-*object functor* $F: \mathbf{K} \to \mathbf{C}$ that on objects sends (X,σ) to X and on morphisms sends $f:(X,\sigma) \to (Y,\tau)$ to $f: X \to Y$. When $\mathbf{C} = \mathbf{Ens}$ this functor is of course nothing but the forgetful functor of Example II.8(1).

Because of this $\mathbf{K} \to \mathbf{K}$ association, it is actually sufficient just to study the usual categories and forget about the 'structured categories.' But the latter tends to put things in a better perspective, and so we shall refer to them occasionally.

EXAMPLES. **Gp** and **Top** can be regarded as categories of sets (i.e., **Ens**-objects) with structure, with $\mathbf{Gp}(X)$ = the set of all group structures on the set X and $\mathbf{Top}(X)$ = the set of all topologies on X. The admissible morphisms are chosen appropriately as group homomorphisms and continuous mappings, repectively.

Topological groups can be regarded as a category of groups with structure; here we have $\mathbf{C} = \mathbf{Gp}$ and $\mathbf{K}(X)$ = the set of all topologies on the group X compatible with the group structure, and a **Gp**-morphism is admissible if and only if it is continuous.

To cite a closer example, the category **N** can be considered as a categories of **S**-objects with structure where $\mathbf{N}((S, F))$ is the collection of all sets of dynamics on the phase space S.

11. Organismic Sets

DEFINITION. An *organismic set* is a finite set S such that

(1) Corresponding to each element $e_i \in S$ there is a set S_i^{a} of *activities* that e_i is capable of performing, and there is a set S_i^{p} of *products* that e_i can make.

(2) The set of all potential activities $S^{\mathrm{a}} = \bigcup_i S_i^{\mathrm{a}}$ and the set of all products $S^{\mathrm{p}} = \bigcup_i S_i^{\mathrm{p}}$ of the organismic set both have cardinality greater than one.

(3) In a given environment E at a given time t, each e_i only exhibits a proper subset $S_i^{\mathrm{a}}(E,t)$ of S_i^{a} and makes only a proper subset $S_i^{\mathrm{p}}(E,t)$ of S_i^{p}. This models *adaptation* and *development* as E and t vary.

(4) S is partitioned into three disjoint subsets S_1, S_2, and S_3 such that $S \sim S_3 = S_1 \cup S_2$ is a "normal" organismic set in itself (i.e., S_3 and its associated S_3^{a} and S_3^{p} are the apparently "useless" parts of the organismic set S.), $S \sim S_2$ is an organismic set that can exist but will not develop, and S_1 is the

core of the organismic set so that $S \sim S_1$ cannot exist (i.e., S_1^a and S_1^p are necessary and sufficient for, at least a short range, existence of S).

(5) Taken alone, i.e., removed from S, each $e_i \in S$ has a *survival time* t_i during which they can exist without the availability of S^p. t_i is short compared to the life span of S.

(6) Elements in S^p (i.e., products) act on various e_i's, and so induce a nonempty set of relations S_R with S. It is these relations that make us recognize an organism or a society as such. Members of S_R are in general k-ary relations with $k \geq 2$.

DEFINITION. An organismic set is of *order n* if its elements are organismic sets of order $n - 1$.

We ascribe the order $n = 1$ to the set of genes of cells. Then a multicellular organism is an organismic set of order $n = 2$. A small group of people (e.g., a family) is an organismic set of order $n = 3$. A tribe, as a collection of interacting families, is of the order $n = 4$, and so on. This leads to the idea of *hierarchy* of organismic sets.

The preceding is a very much simplified version of Rashevsky's original formal definition of organismic sets. Attempts to study organismic sets using category theory were made (see Baianu, 1980, for a list references), but ordinary categories did not seem to be sufficient and the theory of supercategories was developed. We shall simply, for our purpose, consider the picccs we took and see if we can fit them into the formalism of our category **N**.

12. Organismic Category

Let **O** be the category of organismic sets; i.e., an **O**-object is an organismic set S. The elements $e_i \in S$ can be considered as "states" of the system S. Members of the sets S^a and S^p can be considered as "observables" on S. In particular the relations in S_R can be interpreted as the observable-induced relations. Although the real-valued requirement of the observables is not met in this case, one can always "digitize" S^a and S^p and impose artificial numbers on them. Of course this digitization has to be done in such a way that there is a minimal loss of information. This situation is perhaps similar to that of numerical taxonomy when we give numerical values to the different taxa. The activities and products in S^a and S^p can also be considered as dynamics induced on the system. Even the survival time t_i of each e_i can be interpreted as the "inherent dynamics" of the element, or as the bounds (a_{e_i}, b_{e_i}). Thus the map $S \mapsto (S, F = S^a \cup S^p, D = S^a \cup S^p \cup \{t_i\})$ is one on the objects of a functor $\mathbf{O} \to \mathbf{N}$. The various other properties of S can be looked upon as further structures on the objects; in other words, **O** can be studied as a category of **N**-objects with structure.

What, then, are the admissible **N**-morphisms ϕ between organismic sets S and S'? Besides the usual requirements of being an **N**-morphism (which implies $\phi: S \to S'$, $S_i^{\mathrm{a}} \to S_i'^{\mathrm{a}}$, $S_i^{\mathrm{p}} \to S_i'^{\mathrm{p}}$, and $\{t_i\} \to \{t_i'\}$), we would like it to preserve the other structures as well. So we want $\phi(S_i^{\mathrm{a}}(E, t)) \subset S_i'^{\mathrm{a}}(E, t)$, $\phi(S_i^{\mathrm{p}}(E, t)) \subset S_i'^{\mathrm{p}}(E, t)$, and $\phi: S_1 \to S_1'$, $S_2 \to S_2'$, $S_3 \to S_3'$, etc. With such a definition, **O**-isomorphic objects would then be abstractly identical organismic sets.

13. Specialized Subsets and Hierarchy

Admittedly, the previous section contains many hand-waving arguments. But the **O** $\to$ **N** association looks rather promising, and the **O**-morphisms do indeed look like the transformations between abstract diagrams representing biological systems, i.e., the biotopological mappings between graphs of organisms (cf. Rashevsky, 1960). The preceding should be able to acquire mathematical rigor upon further "hard" analysis.

Here we shall just look at two more points. First, the preservation of the "specialized subsets" $S_1, S_2, S_3, S^{\mathrm{a}}$, and S^{p} of organismic sets by **O**-morphisms is rather interesting. Recalling that these five sets represent the core, the developer, the nonessential part, the activities, and the products of S, respectively, we can view **O**-morphisms as mappings of different biological functions. Second, when we have a chain of organismic sets $S^1 \to S^2 \to S^3 \to \cdots$ where S^n is of order n, the arrows (ie., **O**-morphisms) in between them are then mappings from one "level" to the next in the hierarchy of organismic sets. These two ideas of "specialized subsets" and "hierarchy" are explored in detail in Miller's (1978) theory of living systems, our next topic.

14. Living Systems

The general living systems theory is presented in Miller (1978) in a conceptual setting concerned mainly with "concrete" systems that exist in space-time. It is not a mathematical treatment but it does seem to provide an (exhaustive) catalogue of what we know about biological and social systems (up to 1978).

Miller classifies complex systems that carry out living processes into seven hierarchical levels—cell, organ, organism, group, organization, society, and supranational system (cf. the orders $n = 1$–7 of organismic sets). The central thesis is that living systems at all of these levels are open systems composed of subsystems that process inputs, throughputs, and outputs. There are 19 critical subsystems essential for life, some of which process matter or energy (named suggestively ingestor, distributor, converter, producer, matter–energy storage, extruder, motor, and supporter), some of which process information (namely, input transducer, internal transducer, channel and net, decoder, associator, memory, decoder, encoder, and output transducer), and some of which process all three (reproducer and boundary). Together they make up a living system. For a list of what all these critical subsystems are at each of the seven levels of living systems, refer to Table 13-1 of Miller's book.

What we want to do here is merely to point out how much one can achieve from two simple postulates (that of critical subsystems—specialized subsets—and that of hierarchy) and a large collection of observed data. We can see how close Miller's theory of living systems is to Rashevsky's theory of organismic sets, and hence the former can somehow also be put into the framework of our categorical system theory. Here we probably want to map each of the critical subsystems to corresponding ones between systems, and "hierarchical morphisms" can also be considered.

Finally, let us conclude these few sections on the "conceptual superstructure" of physical, biological, and social systems by saying that the relations among the components of these systems are not put in there by the imagination of the observer, as shepherds idly trace out a scorpion in the stars. These relations are inherent in the systems and are empirically discovered by the scientists. They are there, patterning the coacting reality, regardless of whether they are observed or not. We must pass beyond raw empiricism, beyond the provincialism of the disciplines, and learn from the cooperation of experiments and theories—how theories are created to stimulate further experiments and how experiments are designed to evaluate theories. The better we can formulate a problem, the more we will know about where and what to look for, and the closer we will get to obtaining a complete set of observables.

D. Description Spaces

15. The Response Tensor and Description Space

The concept of the response tensor was originally presented (Richardson, 1980) in the context of irreversible thermodynamics. In that paper a metric algebra based upon the dissipation function associated with a system was introduced and a measure of the aging of the system was derived from this algebra. The mathematical formalism underlying the concepts of the response tensor and the space it spans called the description space was then presented as a phenomenological calculus for complex systems in Richardson, Louie, and Swaminathan (1982).

In these few sections we shall show how these ideas are connected to those of categorical system theory. Let us first have a quick review of the basic definitions and postulates of the phenomenological calculus.

DEFINITION. Let V be the vector space $\mathbb{R}^n$ equipped with the standard inner (dot) product. Let $\mathbf{F}$ be a (contravariant) vector in V and $\mathbf{a}$ be a (covariant) vector in the dual space V^* (the space of all linear functionals on V). Then the *dyad* $\mathbf{R} = \mathbf{aF}$ is a bilinear function from $V^* \times V$ to $\mathbb{R}$ (i.e., $\mathbf{R}$ is in $T_1^1(V)$, the tensor space of type $(1,1)$ over V) defined by $\mathbf{R}(\mathbf{a}', \mathbf{F}') = (\mathbf{a} \cdot \mathbf{a}')(\mathbf{F} \cdot \mathbf{F}')$. A finite linear combination of dyads is called a *dyadic*, whose action on $V^* \times V$ is defined the natural way (linearly on the images); so a dyadic is also a bilinear function from $V^* \times V$ to $\mathbb{R}$, i.e., in $T_1^1(V)$.

It follows from dimensional arguments that the collection of all dyadics is actually the whole of $T_1^1(V)$. In other words, each tensor of type $(1,1)$ has a representation as a dyadic.

DEFINITION. Let $\mathbf{R} = \Sigma_i \mathbf{a}_i \mathbf{F}_i$ and $\mathbf{S} = \Sigma_j \mathbf{b}^j \mathbf{G}_j$ be two dyadics. Their *double dot product* is the real number $\mathbf{R} : \mathbf{S} = \Sigma_i \Sigma_j (\mathbf{a}^i \cdot \mathbf{b}^j).(\mathbf{F}_i \cdot \mathbf{G}_j)$.

One can check that the double dot product is independent of the representation of the dyadics and is in particular independent of the basis chosen for V. Also, it is easy to see that : is a positive definite bilinear form. This gives the following theorem.

THEOREM. $(T_1^1(V), :)$ is an inner product space. □

The phenomenological calculus arising from the mathematical structure of $(T_1^1(V), :)$ is based upon three postulates:

POSTULATE 1. A given system is characterized by a set of vectors $\{\mathbf{a}^i : i = 1, 2, \ldots, m\}$ in (the dual space of) $\mathbb{R}^n$, and this set depends on the physical constitution of the system. As far as describing the dynamic response of the system to the imposition of a set of forces (or more generally, *causes*) $\{\mathbf{F}_i : i = 1, 2, \ldots, m\}$ in $\mathbb{R}^n$, they form a complete set of constitutive parameters. The index i denotes subsystems of the system.

POSTULATE 2. The system dynamics are characterized phenomenologically by the dyadic response tensor $\mathbf{R} = \Sigma_i \mathbf{a}^i \mathbf{F}_i$.

POSTULATE 3. The space spanned by $\mathbf{R}$, i.e., for fixed covariant vectors $\mathbf{a}^1, \ldots, \mathbf{a}^m$ in $V^* = \mathbb{R}^n$, the set $[\mathbf{a}^1, \ldots, \mathbf{a}^m] = \{\mathbf{R} = \Sigma_i \mathbf{a}^i \mathbf{F}_i : \mathbf{F}_i \in V = \mathbb{R}^n\}$, is called the description space, and $\mathbf{R}$ is invariant under coordinate transformations.

Thus the description space $[\mathbf{a}^1, \ldots, \mathbf{a}^m]$ is a linear subspace of $T_1^1(V)$ and hence $([\mathbf{a}^1, \ldots, \mathbf{a}^m], :)$ is an inner product space. The condition $|\mathbf{R}| = (\mathbf{R} : \mathbf{R})^{1/2} \geq 0$ for a response tensor is interpreted as a principle of directionality for dissipation, aging, and in fact all cause-and-effect phenomenologies.

Instead of the Euclidean inner product space $V = (\mathbb{R}^n, \cdot)$, we could have used a general Hilbert space H and all of the mathematics would still go through. In the special case of $H = L^2$ we would have incorporated into the setting the "time-dependence" of the causes $\mathbf{F}(t) \in L^2$ and the "cause-dependence" of the constitutive parameters $\mathbf{a}(\mathbf{F}) \in (L^2)^* = L^2$. This "Hilbert description space" turns out to be a very interesting mathematical object, the study of which is certainly a worthy project.

16. The Description Space as a Natural System

Let S be an open subset of the Euclidean space $V = \mathbb{R}^n$. A linear functional $\mathbf{a}$: $\mathbb{R}^n \to \mathbb{R}$ ($\mathbf{a} \in V^*$) can be considered as a real-valued function, i.e., an observable, on the state space S. So the set of constitutive parameters (or "coordinate vectors") $\{\mathbf{a}^i\}$ of the description space $[\mathbf{a}^1, \ldots, \mathbf{a}^m]$ can be considered as the set of observables on S; i.e., $F = \{\mathbf{a}^i\}$.

Recall [see Section IV.4 and Example V.2(5)] that a vector field $\mathbf{F}: S \to \mathbb{R}^n$ defining an autonomous differential equation

$$dx/dt = \mathbf{F}(x) \tag{1}$$

gives rise to a C^1 (continuously differentiable)-dynamics T where $T(x,t) = y_x(t)$ is the unique solution to (1) satisfying $T(x,0) = y_x(0) = x$. It is interesting to note that the converse also holds, namely, given a C^1-dynamics T: dom $T \subset S \times \mathbb{R} \to S$ (where S and $\mathbb{R}$ have the usual topology), there is associated with it a vector filed and hence an autonomous differential equation. Define $\mathbf{F}: S \to \mathbb{R}^n$ by

$$\mathbf{F}(x) = \frac{d}{dt} T(x,t)|_{t=0} \tag{2}$$

then for $x \in S, \mathbf{F}(x)$ is a vector in $\mathbb{R}^n$, which we can think of as the tangent vector to the T-trajectory $y_x(a_x, b_x)$ at $t = 0$. And it is clear that y_x is the unique solution to the autonomous differential Eq. (1) satisfying the initial condition $y_x(0) = x$. Thus this establishes a correspondence between vector fields $\mathbf{F}$ and C^1-dynamics T on S. The collection of m-tuples of causes (or "components") $\{\mathbf{F}_i\}$ defining response tensors $\mathbf{R} = \Sigma_i \mathbf{a}^i \mathbf{F}_i$ can then be interpreted as C^1-dynamics on the phase space S through this correspondence. This family of C^1-dynamics on S is then considered as the set D for the natural system (S, F, D).

So we have established that a description space over $V = \mathbb{R}^n$ is in fact a special kind of natural systems fully equipped with its sets of observables and dynamics. This embedding $[\mathbf{a}^1, \ldots, \mathbf{a}^m] \mapsto (S, F, D)$ is quite remarkable in that even the physical interpretation of the different corresponding entities coincides. The set of constitutive parameters of a description space and the set of observables of a natural system are both indicators of the *complexity* of the system, and the former set is actually what we observe on a physical system (by Postulate 1 of the previous section). The connection $F \leftrightarrow T$ between V and D is even more transparent: a force field is the time derivative of a dynamics [in the sense of Eq. (2)] in classical physics.

17. The Category of Description Spaces

Now that we have found out that description spaces are representable as **N**-objects, the next natural question to ask is, "What are the **N**-morphisms between description spaces that would preserve all the relevant structures?" We can consider description spaces as **N**-objects with structure and look for admissible **N**-morphisms (see Definition VII.10). But because of the $\mathbf{K} \to \mathbf{K}$ association and the underlying object functor $\mathbf{K} \to \mathbf{C}$ mentioned in Section VII.10, we can simply consider the collection of all description spaces as a usual category equipped with a functor to **N**. This functor takes a description space to its associated **N**-object, and an **N**-morphism between description spaces to one between **N**-objects.

Let us call the category of description spaces **R**, i.e., an **R**-object is a description space $[\mathbf{a}^1, \ldots, \mathbf{a}^m]$ (over the inner product space $V = \mathbb{R}^n$). Since a description space has a "linear" structure, we would like an **N**-morphism between two description spaces $\phi: [\mathbf{a}^1, \ldots, \mathbf{a}^m] \to [\mathbf{b}^1, \ldots, \mathbf{b}^l]$ to be "linear" as well:

$$\phi\left(\sum_i \mathbf{a}^i \mathbf{F}_i\right) = \sum_i \phi(\mathbf{a}^i)\mathbf{F}_i \tag{1}$$

and so it is clear that ϕ is determined upon further restriction of the m images $\phi(\mathbf{a}^1), \ldots, \phi(\mathbf{a}^m) \in \{\mathbf{b}^1, \ldots, \mathbf{b}^l\}$ to satisfy (1). The linearity condition (1) is a restriction of ϕ on the observables. This is analogous to the situation in linear algebra where a linear transformation is uniquely determined by its action on a basis of the domain vector space. So the set of coordinate vectors $\{\mathbf{a}^i\}$ does indeed behave like a basis although it is not a basis in the vector space sense. By virtue of ϕ being an **N**-morphism and the linearity condition (1), if, for example, $\mathbf{a}^3 = c_1\mathbf{a}^1 + c_2\mathbf{a}^2$ for $c_1, c_2 \in \mathbb{R}$, then we must have $\phi(\mathbf{a}^3) = c_1\phi(\mathbf{a}^1) + c_2\phi(\mathbf{a}^2)$. Note that the relation between **R** and **N** is astonishingly similar to that between **Vect** and **Ens**, where **Vect** is the category of vector spaces and linear transformations over a fixed field.

18. Special **R**-*Morphisms*

It is shown in Richardson, Louie, and Swaminathan (1982) that if the coordinate vectors $\{\mathbf{a}^i\}$ span a subspace of V^* of dimension $k(\leq n)$, then the description space $[\mathbf{a}^1, \ldots, \mathbf{a}^m]$ is of dimension kn (over $\mathbb{R}$). Now suppose the coordinate vectors $\{\mathbf{b}^j\}$ of a second description space $[\mathbf{b}^1,\ldots,\mathbf{b}^l]$ also span a subspace of V^* of dimension k, then $[\mathbf{b}^1,\ldots,\mathbf{b}^l]$ is again of dimension kn and so we would expect somehow that the two description spaces are "isomorphic."

Now what is an **R**-isomorphism? Clearly, it has to be an **N**-isomorphism in the first place. So far we have neglected the double dot product on the description spaces. Recalling that a linear transformation between two inner product spaces of the same (finite) dimension is an isomorphism if and only if it preserves inner products, we shall say that an **R**-morphism ϕ *preserves double dot products* if for all response tensors **R** and **S**,

$$\mathbf{R}:\mathbf{S} = \phi(\mathbf{R}):\phi(\mathbf{S}) \tag{2}$$

[Note the two double dot products appearing on the two sides of Eq. (2) are on different description spaces.] Then we shall say that the two description spaces $[\mathbf{a}^1,\ldots,\mathbf{a}^m]$ and $[\mathbf{b}^1,\ldots,\mathbf{b}^l]$ (of the same dimension kn) are **R**-isomorphic if there is an **R**-morphism that is an **N**-isomorphism and preserves double dot products between the two spaces. So under this definition, **R**-isomorphic description spaces are abstractly the same with respect to all of their mathematical structures.

Next, suppose $\phi \in \mathbf{R}([\mathbf{a}^1,\ldots,\mathbf{a}^m], [\mathbf{b}^1,\ldots,\mathbf{b}^l])$ is such that there exists an $\varepsilon > 0$ and for every $\mathbf{R} \in [\mathbf{a}^1,\ldots,\mathbf{a}^m]$, $|\mathbf{R} - \phi(\mathbf{R})| < \varepsilon$. This condition can be roughly stated as $|\mathbf{R}(\mathbf{a}) - \mathbf{R}(\mathbf{b})| < \varepsilon$ in which the notation is self-explanatory. This leads us to the idea of the "distance" between response tensors from different description spaces. It is intuitively clear that the closer two description spaces are to being "identical," the smaller the norm $|\mathbf{R}(\mathbf{a}) - \mathbf{R}(\mathbf{b})|$ will be. And conversely the smaller the norm, the more **R**-isomorphic the two spaces are. So while the condition $|\mathbf{R}| \geq 0$ describes the dissipation (i.e., aging) *within* a system, the condition $|\mathbf{R}(\mathbf{a}) - \mathbf{R}(\mathbf{b})| \geq 0$ allows one to compare the extent of aging *between* two systems. The former depends only on the constitution (i.e., structure) of a system itself, and the latter depends on the morphisms (i.e., on how close they can get to being identities) between systems.

There is, moreover, an alternate description of the intersystem comparision of aging. As usual, **R**-monomorphisms give rise to a partial order on the **R**-objects, where an **R**-monomorphism is some natural analogue of an injective linear transformation and an **N**-monomorphism. Thus $|\mathbf{R}(\mathbf{a}) - \mathbf{R}(\mathbf{b})|$ gives an indication of how close two systems are in age while $\mathbf{R}(\mathbf{a}) \leq \mathbf{R}(\mathbf{b})$ gives an ordering, a directionality to aging. This bears a remarkable resemblance to the aspects of *simultaneity* and *temporal succession* in the concept of time we discussed in Section V.8. Perhaps this is not too surprising. After all, although aging and time are distinct concepts, they do share a lot of characteristics in common. In particular, they are both clocks—aging is an intrinsic clock and time is an extrinsic clock—for natural systems.

19. Recent Developments

The epistemological exploration of the phenomenological calculus has been continued since the 1982 paper. The sequence of additional publications so far consists of Louie, Richardson and Swaminathan (1982), Louie and Richardson (1985), Richardson and Louie (1983), Richardson and Louie (1985), and Louie and Richardson (1985). The phenomenological calculus turns out to be a general algorithm for the synthesis of mathematical representations of complex, highly interacting systems, and the metric structure inherent in the algorithm provides relationship-connecting representations.

Topics discussed under the framework of the phenomenological calculus include recognition processes, duality and invariance, projective representations, irreversible thermodynamics, quantum mechanics, relativity, and information. The list is surprisingly diverse.

VIII. CONCLUSION

And the end of all our exploring
Will be to arrive where we started
And know the place for the first time.

T. S. Eliot, *Little Gidding*

Our investigation of natural systems started with the reciprocity between observables and dynamics, which reflects the duality of the structural and functional aspects of a system. The formal study is based on the encoding of these static and dynamic realities into mathematical objects in the appropriate categories, and above all based on the two fundamental propositions of Section III.1.

Although Bertrand Russell's idea on the *scientific process* provides a framework on which to discuss the modelling relation (see Section VI.2), which is crucial to our treatment, the term "scientific process" is somewhat inadequate. It is true that we do learn from observation and collection of experimental data, that we do try to theorize to explain these data, and that we do correlate between experiments and theories in order to better both. But this is *modelling*, which only forms a *part* of the scientific process. Science is more than that. Especially since Albert Einstein, the idea of *creative imagination* enters science in an essential way. Innovations in science are consequences more of the genius of the human brain than of "trial-and-error" modelling. So if the modelling relation diagrams in Section VI.2 were to represent the scientific process, on them we should perhaps superimpose pictures of the human brain. As the nineteenth-century scientist Baron Justus von Liebig said, "Every property of an object may give, under appropriate circumstances, a key to a locked door; but theory is the master key that opens all doors." And abstract theories not directly constructed from models of the observed phenomena are probably the best keys.

Having said that natural systems are parts of the external world hence real while formal systems are creations of our minds hence abstract (see Section VI.1), we should recognize that the distinctions between reality and abstraction are not that clear. The real can simply be considered as a consequence of the abstract: "So the living and the dead, things animate and inanimate, we dwellers in the world and this world wherein we dwell... are bound alike by physical and mathematical law" (D'Arcy Wentworth Thompson). On the other hand, the abstract can simply be considered as a piece of the real: "I believe that mathematical reality lies outside us, and that our function is to discover or observe it, and that the theorems which we prove and which we describe grandiloquently as our creations are simply our notes of our observations" (Godfrey Harold Hardy). These considerations reinforce one of the points we try to make in this study, namely, that the abstract properties of measurement pertain as much to pure mathematics as they do to natural systems, and conversely, that pure mathematics is in a sense a result of our measurements.

In Section VII we saw how categorical system theory can be applied to analyze biological problems. The exploration of each area turned out to be the study of a structured category on the category of natural systems. We examined these categories individually but we did not really consider the relations among them. Recall that there are, roughly speaking, three levels in the hierarchy within category theory: namely, categories, functors, and natural transformations. We certainly did use categories a lot, and functors were also employed occasionally, but we did not really talk about natural transformations other than just defining them (see the Definition in Section II.11). Since functors are used to "relate" categories and the functors between two categories are the objects of the "functor category" (see Section II.12) in which the morphisms are natural transformations, and since natural transformations are intimately related to the concept of *similarity* [Rosen (1978)], it is certainly worthwhile to look into these further. Also , the categories **Top** and **Vect** appeared in various places in this study, and it would be fruitful to look for the connections between these and our "system categories" in more detail.

In sum, this work represents a formal extension of some of the ideas suggested in Rosen (1978). I have expressed the fundamentals of measurement and representation of natural systems in the setting of the abstract mathematical theory of categories. I believe I have achieved in this work, with perhaps the exception of the last section, a level of mathematical rigour that is recognizable as such. In this Section VII, although the mathematics is not as stiff, I have managed to weave together several theories of natural (alias biological, aging, dissipative, organismic, living, complex...) systems from the standpoint of our categorical system theory. So this categorical extension of Rosen (1978) does indeed look promising. This extension is, however, not unique, and there are many areas that can be further explored. These will be some of my projects for the years to come.

Once I came across a passage written by A. Lawrence Lowell; since then it has become the directive of my scientific life:

"You will seek not a near, but a distant, objective, and you will not be satisfied with what you may have done. All that you may achieve or discover you will regard as a fragment of a larger pattern, which from his separate approach every true scholar is striving to descry."

ACKNOWLEDGMENTS

My thanks are due to Professor Robert Rosen, mathematical biologist, my supervisor, teacher, and friend, who taught me, among other things, the importance of pluralism.

I am grateful to Professor I. W. Richardson for his generous support, numerous inspiring discussions, and valuable suggestions on how to improve the manuscript.

I would like to extend my appreciation to my colleagues in the Biomathematics Annex—our Red House—for their friendliness and many helpful comments. Special thanks go to E. A. D. Foster for a critical reading of the first draft of this thesis.

My thanks also go to Professor W. R. S. Sutherland, my advisor from the Department of Mathematics, for his guidance.

I acknowledge the Killam Trust of Dalhousie University for support in the form of an Izaak Walton Killam Memorial Scholarship.

Above all, it is a pleasure to express my deepest gratitude to my wife, Margaret, without whose encouragement and understanding nothing could have been done.

REFERENCES

Arbib, M. A. (1972). Automata theory in the context of theoretical embryology. "Foundations of Mathematical Biology," Vol. 2, Chap. 3, 141–215. Academic Press, New York.

Baianu, I. C. (1980). Natural transformations of organismic structures. *Bull. Math. Biol.* **42**, 431–446.

Belinfante, F. J. (1973). "A Survey of Hidden-Variable Theories." Pergamon Press, New York.

Berrill, N. J. and Karp, G. (1976). "Development." McGraw-Hill, New York.

Bertalanffy, L. V. (1968). "General System Theory." Braziller, New York.

Bohm, D. (1957). "Causality and Chance in Modern Physics." Van Nostrand-Rheinhold, Princeton, New Jersey.

Comorosan, S. (1976). Biological observables. *Prog. Theor. Biol.* **4**, 161–204.

Crick, F. H. C. (1977). Developmental biology "The Encyclopaedia of Ignorance," 299–303. Pergamon Press, London.

Eddington, A. S. (1939) "The Philosophy of Physical Science." Cambridge Univ. Press, London.

Eddington, A. S. (1949) "Fundamental Theory." Cambridge Univ. Press, London and New York.

Eilenberg, S. and Mac Lane, S. (1945) General theory of natural equivalences. *Trans. Am. Math. Soc.* **58**, 231–294.

Güttinger, W. and Eikemeier, H. ed. (1979). "Structural Stability in Physics." Springer-Verlag, Berlin and New York.

Hájek, O. (1968). "Dynamical Systems in the Plane." Academic Press, London.
Haken, H. (1978). "Synergetics: An Introduction," 2nd ed. Springer-Verlag, Berlin and New York.
Jeans, J. (1945). "The Astronomical Horizon." Oxford Univ. Press, London and New York.
Louie, A. H., and Richardson, I. W. (1983). *Math. Modelling* **4**, 555–565.
Louie, A. H., and Richardson, I. W. (1985). *Math. Modelling* **6**, (in press).
Louie, A. H., Richardson, I. W., and Swaminathan, S. (1982). *J. Theor. Biol.* **94**, 77–93.
Mac Lane, S. (1971). "Categories for the Working Mathematician." Springer-Verlag, Berlin.
Manes, E. G. (1976). "Algebraic Theories." Springer-Verlag, Berlin and New York.
Miller, J. G. (1978). "Living Systems." McGraw-Hill, New York.
Rashevsky, N. (1960). "Mathematical Biophysics." 3rd rev. ed. Dover, New York.
Rashevsky, N. (1972) "Organismic Sets." Mathematical Biology, Inc., Holland, Michigan.
Richardson, I. W. (1980). The metrical structure of aging (dissipative) systems. *J. Theor. Biol.* **85**, 745–756.
Richardson, I. W., and Louie, A. H. (1983). *J. Theor. Biol.* **102**, 199–223.
Richardson, I. W., and Louie, A. H. (1985). *Math. Modelling* **6**, (in press).
Richardson, I. W., and Rosen, R. (1979). Aging and the metrics of time. *J. Theor. Biol.* **79**, 415–423.
Richardson, I. W., Louie, A. H. and Swaminathan, S. (1982). A phenomenological calculus for complex systems. *J. Theor. Biol.* **94**, 61–76.
Rosen, R. (1958). The representation of biological systems from the standpoint of the theory of categories. *Bull. Math. Biophys.* **20**, 317–342.
Rosen, R. (1971). "Dynamical System Theory in Biology." Wiley, New York.
Rosen, R. (1972). Some relational cell models: the metabolic-repair system. "Foundations of Mathematical Biology," Vol. 2, Chap. 4, 217–253. Academic Press, New York.
Rosen, R. (1976). Structural stability, alternate descriptions and information. *J. Theor. Biol.* **63**, 19–31.
Rosen, R. (1978). "Fundamentals of Measurement and Representation of Natural Systems," Elsevier, New York.
Rosen, R. (1980). On control and optimal control in biodynamic systems. *Bull. Math. Biol.* **42**, 889–897.
Rosen, R. (1985) "Anticipatory Systems." Elsevier, New York. In Press.
Russell, B. (1931) "The Scientific Outlook." Norton, New York.
Thom, R. (1975). "Structural Stability and Morphogenesis." Benjamin, Reading, Massachusetts.
Thompson, D. W. (1942). "On Growth and Form," 2nd ed. Cambridge Univ. Press, London.
Trinkaus, J. P. (1969). "Cells into Organs." Prentice-Hall, Englewood Cliffs, New Jersey.

3

Organisms as Causal Systems Which Are Not Mechanisms: An Essay into the Nature of Complexity

ROBERT ROSEN

Department of Physiology and Biophysics
Dalhousie University
Halifax, Nova Scotia

I. INTRODUCTION

The thrust of this essay is that the theory of organisms, and of what we shall call *complex systems* in general, requires a circle of ideas and methods that, from the very outset, depart radically from those taken as axiomatic for the past 300 years.

What we shall conclude can be stated succinctly here at the outset, as follows.

1. Our current systems theory, including all that presently constitutes physics or physical science, deals exclusively with a very special class of systems that I shall call *simple systems* or *mechanisms*.

ISBN 0-12-597280-6

2. Organisms, and many other kinds of material systems, are not mechanisms in this sense. Rather, they belong to a different (and much larger) class of systems, which we shall call *complex*.
3. Thus the relation between contemporary physics and biology is not, as everyone routinely supposes, that of general to particular.
4. To describe complex systems in general, and organisms *a fortiori*, an entirely novel kind of mathematical language is necessary.
5. A simple system can only *approximate* to a complex one, locally and temporarily, just as, e.g., a tangent plane can only approximate to a nonplanar surface locally and temporarily. Thus in a certain sense, a complex system can be regarded as a kind of global limit of its approximating simple subsystems.
6. Complex systems, unlike simple ones, admit a category of final causation, or anticipation, in a perfectly rigorous and nonmystical way.

We shall find many other novel consequences of complexity as we proceed; especially to be mentioned are deeper insights into the nature of "information" and the relation of theory to "experiment."

We thus argue, in effect, that any attempt to deal effectively with the material basis of biological organization forces a revamping of our entire traditional scientific epistemology. As we shall see, this revamping entails a number of dramatic consequences, not only for biology, but also for physics, and indirectly, even mathematics.

II. BIOLOGY AND OTHER SCIENCES

Biology is the linchpin of the sciences. Insofar as organisms are material systems, their remarkable properties stand as a challenge, a reproach, and an inspiration to the sciences of matter (physics and chemistry). Insofar as organisms adapt, perceive, and behave, their properties impinge on our technologies to an ever-increasing extent. And insofar as organisms evolve, develop, constitute communities, and form societies, biology also provides both the material substratum and the metaphorical inspiration for all the social, political, and economic sciences. Thus nothing that happens in any other science can be immaterial to biology; conversely, anything that happens in biology ultimately radiates into every corner of scientific thought.

At root, theoretical biology is concerned with only one great question: What is it about certain material systems that confers upon them the characteristics of life, which makes them living beings? All other problems of biology, both theoretical and "practical," are collateral or subordinate to this central question.

It is a significant fact that, despite generations of trying, there is as yet no list of tests, characteristics, or criteria we can apply to a given material system that can decide whether that system is an organism or not. Stated another way, the decision as to whether a given system is an organism is entirely a subjective, intuitive one, based on criteria that have so far resisted formalization, or even articulation. Thus from a strictly rigorous point of view, the subject matter of the science of biology is *undefined*; it is based entirely on an informal consensus essentially akin to pattern recognition, but that consensus is one we all share to a startling degree.

The problem of "What is life?", to use Schrödinger's phrase, first became acute with the triumph of the Newtonian revolution (about which we shall have much more to say subsequently), for one of the major consequences of the Newtonian world view was the obliteration of any distinction between the organic and inorganic. In this view, every material system could be analyzed down to a population of structureless particles moving in fields of force; in principle, the dynamical equations governing any such population could be written down and solved. Among other things, this picture has become the canonical one for scientific *explanation*; therefore the revolutionary developments in physics that created the problem of "what is life," by obliterating any distinction between living and nonliving, also proclaim themselves the only place to look in trying to solve the problem.

Indeed, insofar as contemporary physics claims to deal with material reality in all its manifestations and insofar as organisms are material systems, it is most natural to seek insight into organic phenomena via *biophysics*. Therefore we shall briefly review the present status of this endeavor.

First, we must remark that, historically, the relation between theoretical physics and biology has never been close. None of the great names of physical science, from Newton to the present, have known or cared much about the properties of organisms, and therefore organic phenomena played no essential part in their science (leaving aside such diversions as Maxwell's Demon, Schrödinger's informal essays, and the like). For the past 300 years, the theoretical physicist has exclusively concerned himself with the formulation of *universal laws*. From his perspective biology is concerned with a small class of very special (indeed, *inordinately* special) systems, clearly not the place to look when seeking universal laws. In this light, what makes organisms so special is not their immunity from universal laws of physics, but the apparent multitude of initial and boundary conditions that must be determined in order to bring the laws to bear upon them. The specification of these conditions is regarded by the physicist as an empirical job, and someone else's job at that. In a word, the physicist has always believed that there is no new physics to be learned from organisms; he has never doubted that he deals in the general and that biology concerns only the particular.

The modern biologist has all too avidly embraced this perspective. In a historical sense, the past 50 years have seen biology finally catch up with the

Newtonian revolution that swept over the rest of the scientific world in the eighteenth century. The three-century lag arose because biology had no analogue of the solar system; no way to make immediate contact with the Newtonian ideas. Not until physics and chemistry had elaborated the technical means to isolate and manipulate minute quantities of matter (including organic matter) in the 1930s could one think of a particulate, mechanical basis for biology, i.e., of a *molecular* biology.

The preceding considerations do not represent solely my own subjective assessment. It is worth digressing for a moment to address this further, because the current state of biological science is not generally seen as the culmination of historical trends going back to Newtonian mechanics, or even further back to Descartes. Rather, it is regarded as objective and axiomatic, even though, as we shall see, its diverse historical roots imbue it with obvious paradoxical properties.

In 1970 there appeared a volume entitled "Biology and the Future of Man," edited by Philip Handler, then president of the National Academy of Sciences of the U.S.A. The book went to extraordinary lengths to assure the reader that it spoke for biology as a science, that in it biologists spoke with essentially one voice. For instance, it was emphasized that the volume was prepared not as a mere academic exercise, but for serious pragmatic purposes:

> Some years ago, the Committee on Science and Public Policy of the National Academy of Sciences embarked on a series of "surveys" of the scientific disciplines. Each survey was to commence with an appraisal of the "state of the art".... In addition, the survey was to assess the nature and strength of our national apparatus for continuing attack on those major problems, e.g., the numbers and types of laboratories, the number of scientists in the field, the number of students, the funds available and their sources, and the major equipment being utilized. Finally, each survey was to undertake a projection of future needs for the national support of the discipline in question to assure that our national effort in this regard is optimally productive.... [p. v.]

To address such serious matters, we are then told that the academy proceeded as follows:

> Panels of distinguished scientists were assigned subjects.... Each panel was given a general charge...as follows:
>
> The prime task of each Panel is to provide a pithy summary of the status of the specific sub-field of science which has been assigned. This should be a clear statement of the prime scientific problems and the major questions currently confronting investigators in the field. Included should be an indication of the manner in which these problems are being attacked and how these approaches may change within the foreseeable future. What trends can be visualized for tomorrow? What lines of investigation are likely to subside? Which may be expected to advance and assume greater importance?... Are the questions themselves...likely to change significantly?... Having stated the major questions and problems, how close are we to the answers? The sum of these discussions, panel by panel, should constitute the equivalent of a complete overview of the highlights of current understanding of the Life Sciences. [p. vi.]

There were 21 such panels established, spanning the complete gamut of biological sciences and the biotechnologies. The recruitment for these panels consisted of well over 100 eminent and influential biologists, mostly members of the academy. How the panelists themselves were chosen is not indicated, but there is no doubt that they constituted an authoritative group.

In due course, the panels presented their reports. How they were dealt with is described in vivid terms:

In a gruelling one week session of the Survey Committee... each report was *mercilessly* exposed to the criticism of all the other members.... Each report was then rewritten and subjected to the *searching*, sometimes *scathing*, criticisms of the members of the parent Committee on Science and Public Policy. The reports were again revised in the light of this exercise. Finally, the Chairman of the Survey Committee... devoted the summer of 1968 to the final editing and revising of the final work. [p. vii.]

Thus we have good grounds for regarding the contents of this volume as constituting a true authoritative consensus, at least as of 1970. There are no minority reports; no demurrals; biology does indeed seem guaranteed here to speak with one voice.

What does that voice say? Here are a few characteristic excerpts:

The theme of this presentation is that life can be understood in terms of the laws that govern and the phenomena that characterize the inanimate, physical universe and, indeed, that at its essence life can be understood *only* [emphasis added] in the language of chemistry. [p. 3.]

A little further along, we find this:

Until the laws of physics and chemistry had been elucidated, *it was not possible even to formulate* [emphasis added] the important, penetrating questions concerning the nature of life.... The endeavors of thousands of life scientists... have gone far to document the thesis... [that] living phenomena are indeed intelligible in physical terms. And although much remains to be learned and understood, and the details of many processes remain elusive, those engaged in such studies hold *no doubt* [emphasis added] that answers will be forthcoming in the reasonably near future. Indeed, *only two major questions* [emphasis added] remain enshrouded in a cloak of *not quite* [emphasis added] fathomable mystery: (1) the origin of life...and (2) the mind-body problem... yet (the extent to which biology is understood) even now constitutes a satisfying and exciting tale. [p. 3.]

Still further along, we find things like this:

While *glorying* [emphasis added] in how far we have come, these chapters also reveal how large is the task that lies ahead.... If [molecular biology] is exploited with vigor and understanding... a shining, hopeful future lies ahead. [p. 6.]

And this:

Molecular biology provides the closest insight man has yet obtained of the nature of life—and therefore, of himself. [p. 64.]

And this:

> It will be evident that the huge intellectual triumph of the past decade will, in all likelihood, be surpassed tomorrow—and to the everlasting benefit of mankind. [p. 130.]

It is clear from such rhapsodies that the consensus reported in this volume is not only or even mainly a scientific one; it is an emotional and aesthetic one. Indeed, anyone familiar with the writings of Newton's contemporaries and successors will recognize them.

The volume to which we have alluded was published in 1970. But it is most significant that nothing fundamental has changed since then.

Despite this overwhelming commitment to the mechanical, there is at the same time another thread running through contemporary biology, one quite incompatible with mechanics. A good statement of this was given by Jacques Monod (1971):

> We can assert today that a universal theory, however completely successful in other domains, could never encompass the biosphere, its structure and its evolution as phenomena deducible from first principles.... The thesis I shall present... is that the biosphere does not contain a predictable class of objects or events but constitutes a particular occurrence, compatible with first principles but not deducible from these principles, and therefore *essentially unpredictable* [emphasis added]. [p. 42.]

In other words, the important features of organisms are the result of accidents, and hence governed by no laws at all. Biology thereby becomes a branch of history, not of science. All that can be said for this peculiar combination of mechanism and historical accident is that it allows the modern biologist the luxury of enjoying the fruits of both mechanism and vitalism with an unsullied conscience.

As a matter of fact, there is not a shred of evidence supporting any of this received picture. Quite the contrary; as we proceed, we shall see much that is totally incompatible with it. The reader may take my word for it that a great deal of additional contrary evidence could be adduced with ease. Such incompatibilities are simply ignored by most biologists, and by most physicists as well, on the grounds that merely the acquisition of more information, more data, will somehow resolve them. In their view, what fault there is is not a fault of principle but of practice.

Indeed, it is certainly true that if the fault is one of principle, then the fault is something inherent in the mechanical paradigm itself, something present from the outset. To find it, we need to go back to the very beginnings, to the basic epistemological presuppositions that have become so familiar and axiomatic that to challenge them seems today tantamount to challenging science itself. However, this is exactly what we shall do. To make it clear that we are not operating in a vacuum, we shall preface the more specific analysis with a concrete example of an approach that takes us immediately out of the Newtonian, mechanical context. This is the approach to organisms first systematically developed by Nicolas Rashevsky in 1954 and termed by him *relational biology*.

The upshot of these developments, to be reported later, is thus very much in the spirit of a pregnant remark that Einstein is supposed to have made to Leo Szilard: "One can best appreciate, from a study of living things, how primitive physics still is."

III. RELATIONAL BIOLOGY

As noted earlier, the term *relational biology* was coined by Nicolas Rashevsky in 1954 to distinguish it from more familiar approaches, which he characterized as *metric biology*; Rashevsky, who had himself been trained as a theoretical physicist, was the great pioneer in the application of quantitative physical ideas to the elucidation of the material basis of organic behavior. He almost singlehandedly created the field he called "mathematical biophysics," which, in his words, would stand in the same relation to experimental biology as mathematical physics stands to experimental physics.

In modern terms, Rashevsky's initial approach was mechanist and reductionist. His earliest work involved those areas in which physical processes and biological behavior naturally seemed to intersect—the physical basis of cell division (cytokinesis) and reaction–diffusion processes generally (this was in 1930); nerve conduction and nerve excitation, leading to the first network theories of the central nervous system and the brain; the control, form, and dynamics of the cardiovascular system, and many others.

By 1950 the feasibility and fecundity of mathematical modelling as a probe of biological phenomena was well established, owing largely to Rashevsky's own work and that of his school. But Rashevsky himself was beginning to feel dissatisfied. The source of this dissatisfaction lay in the fact that, in the process of modelling its individual features, the organisms *qua* organism seemed to have disappeared. He began to wonder where and why it had disappeared, and how it could be retrieved. Thus in a spontaneous way he began to struggle with his own reductionism.

The answer he came up with, in his pioneering paper of 1954 ("Topology and Life"), was roughly as follows. Heretofore, we have supposed that we can resolve an organism into *physical* subsystems, understand each of these in detail through traditional modes of physical and mathematical investigation, and when we are done, the original biological organization to which these material subsystems belonged will reemerge *of itself*, as a *consequence* of the nature of these subsystems separately. This is still, of course, the reductionistic credo, most firmly ensconced in the field of molecular biology, which hardly existed in Rashevsky's time. It is a modern version of the idea that any *mixture* (i.e., a system of many phases) must be resolved into its constituent *pure phases* and that the properties of the original mixture may then be *inferred* from those of its constituent pure phases, as logical consequences thereof.

However, as Rashevsky realized better than anyone else, this did not seem to be happening; quite the contrary. Moreover, so many utterly diverse kinds of *physical* systems could be organisms that it became a really serious problem to try to understand, on a *physical* level, how all of them in their physical diversity could manifest those commonalities of behavior and organization that make us recognize them as being alive. That is, Rashevsky grasped that what he called the "perceived unity of the organic world" could not, either conceptually or technically, be approached directly through the "metric" approaches he (and everyone else) was then using.

He thus proposed something extremely radical. In effect, he said something like this: We are really interested in the organizational features common to all living systems; and in their material structures only insofar as they support or manifest those features. Therefore, we have heretofore approached organisms in precisely the wrong way; we have abstracted out, or thrown away, all those global organizational features in which we are really interested, leaving ourselves with a purely material system that we have studied by purely material methods, hoping ultimately to recapture the organization from our material studies. This has not happened. Why do we not then *start* with the organization? Why do we not, in effect, *abstract away the physics and the chemistry*, leaving us with a pure organization, which we can formalize and study in completely general abstract terms; and recapture the *physics* later through a process of *realization*? It was basically this endeavor that he called relational biology.

His initial approach was a simple one; he represented biological organization through directed graphs, whose nodes were "biological functions" and whose directed edges were relations of temporal or logical precedence. His notion of an "abstract biology" was a class of such graphs, all canonically generated from an initial "primordial graph," and hence all canonically related through graph morphisms of special types. These "abstract biologies" had many striking properties; e.g., from suitable knowledge of any graph in such an abstract biology, one could reconstruct the primordial and hence the whole biology.

In his relational biology, then, Rashevsky anticipated many more recent mathematical ideas. However, the time was quite wrong for his new relational ideas to find any acceptance anywhere. In biology, the "golden age" of molecular biology was just beginning; experimentalists had no time or use for anything of this kind. Those who considered themselves theorists either were preoccupied with the reductionistic modelling that Rashevsky had earlier taught them or were bemused by seductive ideas of "information theory," games theory, cybernetics, and the like, regarded Rashevsky and his ideas as generally archaic because he did not take direct cognizance of their enthusiasms.

Thus his ideas about relational biology ended in Limbo, unread or forgotten. However, what he had done was, in effect, to propose a whole new way of representing material reality; a way that, when pursued, pointed to glaring defects and omissions in the older, received ways. We shall now see how this comes about, in a rather different and simpler relational context than Rashevsky's original one.

IV. THE (M, R)-SYSTEMS

The (M, R)-systems comprise a class of relational cell models, first proposed by me in 1958. As with any relational approach, the problem is to try to characterize at least some fundamental organizational feature common to all cells, independent of their specific physicochemical structures. In the (M, R)-systems, the organizational feature taken as central is the distinction between "cytoplasm" and "nucleus." In traditional cytology, the cytoplasm is regarded as the seat of metabolic activity, the province of enzymes that process environmentally derived materials and convert them into metabolically important new forms. The nucleus, on the other hand, is the seat of the genome and in particular plays the guiding role in the synthesis of the metabolic machinery that sits in the cytoplasm. The (M, R)-systems were invented as the simplest and most general class of mathematical systems that embodies this kind of organization. And whatever else a real cell may be, it must also be (or better, must also realize) an (M, R)-system.

This organization may be seen in the simplest imaginable (M, R)-system, which can be expressed as

$$A \xrightarrow{f} B \xrightarrow{\Phi} H(A, B) \tag{1}$$

Here f is simply a mapping from a set A of *environmental inputs* to a new set B of *environment outputs*. As noted earlier, this is the abstract equivalent of "cytoplasm." The mapping

$$\Phi: B \to H(A, B)$$

is the abstract counterpart of "nucleus"; in effect, it takes the outputs of metabolic activity in the cytoplasm and converts it into more metabolic machinery. The range of Φ is thus a set of *mappings*, as indicated.

Mathematically, then, an (M, R)-system is simply a collection of mappings, together with their domains and codomains. The theory of such systems is a part of category theory, and I believe that the (M, R)-systems provided the first indication of the deep role that category theory could play in theoretical science. Indeed, the (M, R)-systems themselves form a category whenever the category from which their sets and mappings are drawn is specified.

The (M, R)-systems, though simple in concept, possess a remarkably rich theory. Let us indicate one important property. As matters stand, we have embodied in the (M, R)-system only one of the characteristics traditionally associated with the cell nucleus; namely, its role in synthesis of the cellular metabolic machinery. But any nucleus worthy of the name has one other decisive property; it *replicates*. In all other cell models known to me, replication must be superimposed as an additional *ad hoc* property. But in (M, R)-systems, there are replication mechanisms inherent in the organizational features already represented; requiring only a further mathematical condition to be satisfied and no further *ad hoc* hypotheses.

Let us consider the simplest situation. Quite generally, let X, Y be any sets and let $H(X, Y)$ be a set of mappings from X to Y. It is well known that every element $x \in x$ induces a mapping

$$\hat{x}: H(X, Y) \to Y$$

by writing

$$\hat{x}(f) = f(x).$$

This map $\hat{x}$ is often called an evaluation map; such maps are familiar, for instance, in linear algebra, where they identify a vector space with its second dual space. In particular, let us put

$$X = B, \qquad Y = H(A, B).$$

Then we can regard each $b \in B$ as inducing a map

$$\hat{b}: H(B, H(A, B)) \to H(A, B).$$

If this map $\hat{b}$ is *invertible*, then its inverse is a map

$$\hat{b}^{-1}: H(A, B) \to H(B, H(A, B)).$$

Now look at the original (M, R)-system. The repair map Φ is obviously an element in $H(B, H(A, B))$. Thus if $a \in A$, $b = f(a)$, and $\hat{b}^{-1}$ exists, we can extend the original (M, R)-system as follows:

$$A \xrightarrow{f} B \xrightarrow{\Phi} H(A, B) \xrightarrow{\hat{b}^{-1}} H(B, H(A, B)). \tag{2}$$

The new map added to the right-hand side is a *replication map*, and by its very definition, it clearly replicates the "nuclear" mapping Φ.

However, there is a condition to be satisfied, namely, that the evaluation map $\hat{b}$ be invertible. This means

$$\hat{b}(\Phi_1) = \hat{b}(\Phi_2) \qquad \text{implies} \qquad \Phi_1 = \Phi_2$$

or

$$\Phi_1(b) = \Phi_2(b) \qquad \text{implies} \qquad \Phi_1 = \Phi_2.$$

This latter is a condition on the original *category*; it is reminiscent of a unique-trajectory property, or biologically, of a one-gene, one-enzyme hypothesis.

It should also be noted that the final two mappings in Eq. (2) themselves consititute an (M, R)-system, in which Φ, the "nuclear" map for the original system, is now a "cytoplasmic" map, and $\hat{b}^{-1}$, the replication map for the original system, is now the "nuclear" map for the new one. Thus *mathematically* there is no intrinsic difference between these processes; any map in our category could be realized either as a metabolic map, a repair map, or a replication map, depending entirely upon the *context*.

Thus from just this quick overview we can see that the (M, R)-systems possess novel characteristics, bearing on biology in a unique way. But as always when one attempts to do theory, one confronts a banal but unavoidable question: *Is it testable*, and if so, *how*? We have been brought up to believe that a theory that is not testable (i.e., falsifiable) is worthless. And indeed, it is also considered part of the *theorist's* job to make theory testable in this sense, in effect to construct some kind of experimental protocol for this purpose, even if only in principle.

It was this exigency that first forced me into epistemology, for I was convinced that a relational description of an organism is as valid, as *physical*, a description as any conventional physicochemical one. But, like the wave functions of quantum mechanics, it is a description pertaining to a *class* of physically diverse (though functionally equivalent) systems; and as long as *experimental test* exclusively means verifying some kind of specific physicochemical operation on individual systems in such a class, there was in principle no way that the relational descriptions could in fact be tested in the conventional sense. For as we have noted, it is precisely such physicochemical particulars that are abstracted away in the process of generating the relational model.

It should be noted that relational approaches such as the one we are discussing do allow us, even as they stand, to make assertions about the physicochemical nature of biological organizations, but these assertions are all negative ones (i.e., about what cannot happen) and thus are considered empirically unacceptable. For instance, I was able to show that any attempt to add a new component (either metabolic or repair) to a given (M, R)-system would generally destroy the overall pattern of cellular organization, that is, would not result in a new (M, R)-system. In fact, to construct a new (M, R)-system that contains a given one and a new component is not an easy matter. The situation is somewhat analogous to attempting to add a new instruction to a computer program; this generally requires a whole subroutine, which may be much larger than the original program was. Such considerations turn out to have obvious implications for genetic engineering, whose practitioners are presently finding out about such limitations the hard way.

In any case, the most obvious way of making contact with the conventional universe of physicochemical descriptions and hence of generating predictions testable by conventional physicochemical experimental techniques (which, as we have noted, is what *testable* means) is through the process of *realization* referred to earlier. For such a realization must, on the one hand, have the relational features of an (M, R)-system and, on the other hand, be a conventional description of a specific physicochemical system and thus be amenable to traditional notions of testability. In particular, we would seek realizations for which *replication maps are also realizable*, for it is in connection with these that the most novel and interesting predictions can be made. In fact, the successful realization of such an (M, R)-system is tantamount to the synthesis of a novel, autonomous life form.

The strategy to be followed in physically realizing an abstract organizational structure like an (M, R)-system seemed at first to me not too different from that followed by an engineer in designing a real physical structure to meet some given initial set of functional specifications. For here, too, we must reach into a class of physically diverse but functionally similar systems and pick one out. The usual criterion for this selection purpose is one of *optimality* (e.g., least cost). Indeed, I might assert that optimality is the *canonical* way of selecting individual elements from equivalence classes; one may think even of such things as the Jordan canonical form of ordinary matrices (in which the number of 0 entries is maximized).

But the problem did not turn out to be that straightforward after all. The process of determining why it was not is in fact the main purpose of this discussion, and it is this to which we now turn.

V. A FIRST ATTEMPT AT REALIZATION OF (M, R)-SYSTEMS

Let us begin by reviewing an early attempt of mine (1964) to solve the realization problem. It seemed to me that a first step would be to transform *mathematically* the (M, R)-systems to a form in which the various sets and mappings of the (M, R)-system could be interpreted in terms of the states of some system and a set of dynamical laws could be superimposed thereon. This was at least the conventional language in which physical systems were universally described; hence realizing this kind of mathematical object would be much easier than realizing an (M, R)-system directly.

The first idea that came to mind was the language of sequential machines, or finite automata. This is in effect the language of classical dynamical system theory (or better, of control theory) paraphrased to the constraints of discrete time and discrete states. Furthermore, it is closely associated with certain kinds of material systems that are important in biology and elsewhere; in biology, in the representation of neural nets (brains) and genetic control in cells (operon networks); in the general theory of digital computing devices; and elsewhere, including the basic theory of mathematics itself.

In its most general formal terms, a sequential machine is a 5-tuple $(A, B, S, \delta, \lambda)$ consisting of three (finite) sets and two mappings:

1. an *input set* or *input alphabet* A;
2. an *output set* or *output alphabet* B;
3. a set of (internal) *states* S;
4. a *next-state* map, $\delta: S \times A \to S$;
5. an *output map*, $\lambda: S \times A \to B$.

The "sequential" aspect comes from iterations of the next-state map and output map, through which we allow the machine to operate on *strings* $\omega = a_{i_1}, \ldots, a_{i_n}$ of elements of A (i.e., elements of the free monoid $A^{\#}$ generated by A) instead of on A itself; the kth element a_{i_k} of a word $\omega \in A^{\#}$ is then thought of as the one presented to the machine at the kth *instant*; and the indices themselves thus pertain to the instants of a discrete time. It is well known that such devices can themselves always be *realized* by neural nets in many ways.

Now an abstract (M, R)-system can itself be realized by something like a sequential machine. In fact, if we consider the (M, R)-system of Eq. (1), we can define

1. input set $= A$;
2. output set $= B$;
3. state set $S = H(A, B)$;
4. next-state map $\delta: A \times S \to S$; defined by $\delta(a, f) \equiv \Phi(f(a))$;
5. output map $\lambda: A \times S \to B$; defined by $\lambda(a, f) \equiv f(a)$.

At first, this looked extremely promising. Biologically, there were a host of network realizations now available (e.g., operon networks). Mathematically, there were a number of possibilities for passing from discrete to continuous time, i.e., to true dynamical and control systems, and thence to explicit "hardware" realizations, which would comprise "cells" of perhaps utterly novel kinds. In these various realizations, one could explicitly seek those situations in which replication mappings were realizable, for these were of the greatest interest.

There were indeed many interesting conclusions that could be drawn from just these possibilities. But the really fundamental problems remained refractory to this whole approach. In a nutshell, the reason lay in the mathematical dichotomy between *set* (object) and *mapping* in the (M, R)-system. In a network realization, a "state" of the network is a pattern of activation in the elements that constitute the network, while the "next-state mapping" is embodied in the wiring diagram of the network. But intuitively, in the (M, R)-system, both the metabolic map(s) f and the nuclear or repair maps Φ should themselves be embodied in (or realized by) physical structures, and their mapping properties should be a consequence of these structures. When we realize $\Phi(f(a))$, for example, this is abstractly a *mapping* ($f\colon A \to B$) in the (M, R)-system; it is a pattern of excitation (i.e., a single state) in a network; but it should be a material structure in the kind of realization we are actually seeking. Even more, the map Φ itself in the (M, R)-system is a wiring diagram in a network realization, a pattern of specificities in an operon network, but, in fact, it should be realized itself as a material *structure*, from which all these mapping properties should follow.

These considerations led to a fundamental rethinking of the whole idea of how to go about realizing any kind of abstract relational description of a material system, and thence to the whole problem of trying to invert the process by which any kind of mathematical description, or model, of a material system is obtained in the first place. In particular, the very fact that the same mathematical formalism (e.g., a network) could be *interpreted* in so many disparate physical ways ultimately led me to suspect that something crucial might be missing *from the mathematics itself*. In other words, I began to entertain the possibility that our conventional mathematical descriptions of physical reality, which have essentially gone unquestioned for three centuries, might themselves be fundamentally deficient, that it was this deficiency that was responsible for the problems posed by an attempt to realize physically an abstract functional organization.

Let us then briefly review what is involved in the way we conventionally build mathematical or theoretical pictures of the material world. As we shall soon see, the procedure involves *several* tacit hypotheses about the material world, which show up most clearly when we attempt to invert the procedure and *realize* a description. These hypotheses may not be true; I would now argue that they are not true in general. Removing them leads us, in fact, to a whole new epistemology, with all the attendant implications of that fact. We will describe all this in the subsequent sections, beginning with a brief description of the nature of our current ideas regarding the imaging of material reality and some of their basic historical roots.

VI. THE MODELLING RELATION

All of our thought processes, from the most mundane aspects of daily life to the deepest reflections of theoretical science, are based on two parallel postulates.

First is the belief that the sequence of events that we perceive in the external world is not entirely whimsical, chaotic, and arbitrary but obeys definite laws or relations. The relations that exist between events in the external world, and that govern their succession, collectively constitute what we call *causality*. Without a belief in causal order, there could be no science and, very probably, no sanity.

But a belief in a causal order relating events in the external world is only one part of the story. The other part is an independent belief that this causal order relating events can be, at least in part, grasped and articulated by the mind. It is a belief that, in some deep sense, the causal order relating events can be mirrored in a corresponding relation between propositions that describe these events. Now such propositions belong to an internal, symbolic, linguistic world and hence cannot themselves be related by any kind of "causality." But there is another kind of order through which propositions can be related, and that is a *logical* order or *implication.*

Thus we must also believe that the causal order, relating events in the external world, can be brought into congruence with a logical or implicative order in some appropriate logical, symbolic world of propositions describing these events. When such a congruence is established, *implications* in the logical system become *predictions* about the causal order.

These two beliefs together constitute the idea of natural law. It is the entire task of theoretical science to establish the congruence between causal order in the external world and implicative order in the formal world, which embodies the very idea of natural law.

The preceding remarks can be summed up succinctly in a diagram:

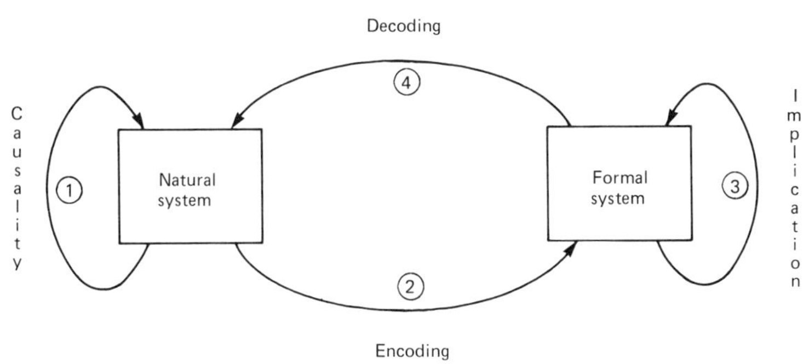

FIG. 1

(cf. Chapter 2). In this diagram the left-hand box represents the external world or some fraction thereof. The sequence of events it exhibits is governed by causal relations, as represented by arrow 1. On the right-hand side sits some formal system, whose elements are governed by relations of implication or logic (arrow 3). We may establish relations between these two diverse worlds by means of "encodings" (arrow 2), whereby attributes of the external world are identified with, or named by, corresponding elements in the formal system; and by "decodings" (arrow 4), whereby elements of the formal system are treated as names of, or symbols for, attributes of events.

We shall say that a *modelling relation* has been established between the natural system and the formal system when the preceding diagram of arrows *commutes*; i.e., when

$$\text{arrow } 1 = \text{arrows } 2 + 3 + 4.$$

In this case, one *always* obtains the same answer; whether one simply sits as an observer and watches the causal order unfold in the natural system (arrow 1) or whether one encodes attributes of that system as propositions (i.e., initial conditions or hypotheses) in the formal system (arrow 2), generates new propositions from these through the inferential structure (arrow 3), and then decodes these back into assertions or predictions (arrow 4).

If a modelling relation exists between a natural system N and a formal system F, we may call F a *model* of N, or N a *realization* of F.

Since mathematics, in the broadest sense, is the study of implication relations in formal systems, or the art of extracting conclusions from premises, it follows that mathematics is integrally involved in the study of natural law. Indeed, many of the deepest questions of theoretical science are concerned with specifying the kinds of formal or mathematical systems that can sit on the right-hand side of the preceding diagram and the kinds of mathematical relations that can exist between them. For instance, the whole problem of *reductionism* involves nothing else.

The equally important inverse or dual problem [i.e. given a formal system on the right-hand side of such a diagram, to determine the class of *natural* systems that can realize it, and the relations that exist between them (variously called *analogy*, *similarity*, or *scaling* relations)] has received no study commensurate with its importance.

We shall be concerned with *both* the class of formal systems that can be put into a modelling relation with a given natural system and with the class of natural systems that can realize a formal system (model). Clearly, these two classes are closely related and should be considered together. For the remainder of this work, then, we shall consider some general aspects of the modelling relation and its ramifications, with a particular eye on the problem with which we started; namely, the construction of realizations of relational models like the (M, R)-systems. Indeed, in the light of the discussion we have just given, encapsulated in the diagram of Fig. 1, we can see clearly how the very idea of a relational model requires *both* (1) a new look at the mathematical structures that can model a natural system (in our case, a "cell") and the mathematical relations between these models and (2) the class of natural systems that realize one (or more) formal structures.

In the process, we will see what radical profundities Rashevsky really unleashed with his apparently innocent ideas about a "relational biology."

VII. THE NEWTONIAN PARADIGM

In this section we shall briefly review the salient features of the class of mathematical or formal systems that are now accepted as models of material reality (what we have called "natural systems"). That is, these are the formal systems that can sit on the right-hand side of a commutative diagram such as Fig. 1. As we shall see, our basic ideas on this subject go back, in one way or another, essentially unchanged, to the mechanics of Newton's *Principia*. Despite enormous *technical* variations in mathematical language (e.g., from classical to relativistic or quantum; from continuous time to discrete time; from continuous state to discrete state; from deterministic to stochastic; from autonomous to forced; from finite-dimensional to infinite-dimensional; etc.), the basic epistemological presupposition remains the same, untouched and all but unnoticed.

This basic presupposition, as we shall see, is that systems have *states* and that upon these states some kinds of *dynamical laws*, or *equations of motion*, are superimposed. The states represent in a sense what is intrinsic, while the dynamical laws reflect the nature of the impinging environment in acting on what is intrinsic. Thus the dichotomy between states and dynamical laws embodies a distinction between *system* and *environment*. Also, in a formal way, the dualism of states and dynamical laws exactly parallels the purely *mathematical* dichotomy between propositions and inferential laws, or production rules, which is nowadays considered as the anatomical foundation of any mathematical formalism whatsoever.

But this basic presupposition, so familiar and axiomatic to us all, involves tacit hypotheses, not just one but several, about the natural world and its mathematical images. We will examine these in detail in the following sections. For the moment, simply review some of the salient formal and historical roots of what we shall call the *Newtonian paradigm*.

First, it must be recognized that Newton's *Principia* was certainly one of the most influential works of human history. In its own time, it was regarded as the capstone of the Renaissance; the culmination of the rational mind and its power to grasp natural law, a fountainhead of optimism and enlightenment.

In subsequent times, it unleashed successive waves of scientific advance that are still going on. It has set the standards for scientific investigation and for scientific explanation to such a degree that alternatives have become essentially unthinkable. This is equally true for modern developments, like quantum theory, which have transcended some aspects of Newton's original formalism, but still, as we shall see, subscribe to exactly the same epistemological presuppositions.

The influence of the *Principia* has radiated in two distinct directions: a reductionistic direction and a paradigmatic direction. We will consider them in turn.

From a reductionistic point of view, we consider the thrust of the *Principia* as the dynamics of systems of mass points. The Newtonian particles are devoid of internal structure; their only attributes are constitutive parameters like mass, which are time independent, and position, which is time dependent. Indeed, the *only* temporally variable attributes of a Newtonian particle are its position and the temporal derivatives of position. Thus the basic problem of Newtonian mechanics (and indeed the *only* problem associated with systems of Newtonian particles) is to tell where the constituent particles are located at any given instant, i.e., to specify configuration as a function of time.

Newton recognized that the arbitrary specification of configuration at an instant placed no restriction on the velocities of the constituent particles; i.e., both configuration and the first temporal derivative of configuration could be chosen completely arbitrarily. One might think that the same would be true for second time derivatives of configuration (i.e., acceleration) and for all higher time derivatives. But here Newton interposed his deep insight. He said, in effect, that the rest of the world (i.e., the *environment* of our system of mass points) exerts *forces* on the particles. *What* these forces are, intrinsically, cannot be (and need not be) specified, but the *effect* of these forces is to determine the acceleration of the particles of our system. That is, insofar as the "force" experienced by a particle is determined by where it is and how fast it is going, the *acceleration* of the particle, and hence all higher temporal derivatives of configuration, are then completely determined by configuration, rate of change of configuration, and the "forces" then imposed by the rest of the world. *Mathematically*, this amounts to expressing acceleration *recursively* as a function of the *lower* temporal derivatives. This expression is the *dynamical law* governing the system; by a mathematical process of *integration* we can convert this dynamical law into a relation giving *configuration* as a function of *time*.

This beautiful conception has a number of sweeping implications:

(1) Insofar as *any* material system can be resolved into a system of structureless particles, Newtonian mechanics seems to provide a recipe, or algorithm, for the study, modelling, and representation of *any* system. That is,

it in principle enables us to construct the diagram of Fig. 1, for *any* material system on the left-hand side. It establishes simultaneously the nature of the formal, mathematical image *and* the encoding and decoding arrows that convert the formal system into a model.

(2) Once the Newtonian picture is accepted, it becomes a purely *empirical* problem to determine, for a given system of interest, what are its constituent particles and what are the forces imposed on them. Thus in the light of the Newtonian approach, there followed historically an enormous empirical shift. We can in fact directly see this shift, in our own time; molecular biology is essentially the result of the belated percolation of Newtonian concepts directly into biology, something that only became technically feasible within the past 30 or 40 years.

(3) The encoding and decoding arrows in Fig. 1 that Newton posited have become axiomatic, and thus in effect invisible. They are no longer recognized as the pivots on which the whole picture turns, but have become as necessary a part of scientific thought as Euclidean geometry was prior to 1800.

In summary, the mathematical image that Newtonian mechanics gives us of a family of structureless particles acted upon by forces is as follows. Such a system is represented by a manifold of possible *phases* (configurations plus velocities) on which a set of equations of motion, representing *forces*, is superimposed. It must be emphasized that this mathematical image is *not* an abstraction; every shred of physical reality of such a system has a corresponding mathematical image somewhere in this picture and requires only a *technical* mathematical exercise to make it visible. All this, by the way, is completely preserved in the transition to quantum physics; the only novelty (though it is a major one) resides in the replacement of the Newtonian phases by a more general space of states, related in a much more complicated way to what we actually measure.

With this as background, we turn now to a brief consideration of the paradigmatic aspect of the Newtonian conception. This involves the use of the *language* in which Newtonian mechanics is couched, to describe systems that have not been, or perhaps cannot be, described as systems of mass points—ecosystems, economic systems, chemical reactors, and so on. The *states* of such a system are the *analogs* of the Newtonian phases; the dynamical laws the *analogs* of Newtonian forces. In this light, as we have noted earlier, *every* mode of system description known to me is nothing more than a paraphrase or an adaptation of the Newtonian *language*, interpreted by Newton initially in terms of particulate mechanical systems, but now interpreted in a wider context.

These two aspects of Newtonian ideas, the reductionist and paradigmatic, come together in the fundamental reductionist assumption that, among all possible encodings of a natural system, there is a *biggest* one, which maps *effectively* on all the others. If every material system is indeed a system of material particles, this would be the original Newtonian one. Mathematically, this biggest description is like a free object in the set of all encodings of a given system. Traditional reductionist ideas, especially in biology, rest entirely on the posited existence of such a biggest encoding and on the assumption that it indeed maps effectively on every other encoding.

We now remark that the relational models described earlier are not of this Newtonian character; they fall outside the paradigm from the very beginning. We will now turn to a study of what this means, and in the process we will find that the Newtonian encodings, far from being universal, are very special indeed.

VIII. MAPPINGS AND "SYSTEM LAWS" IN THE NEWTONIAN PARADIGM

In the previous sections we posed the problem of realizing relational models like the (M, R)-systems. The basic ingredients of such relational models are abstract sets and mappings between them. We take up that discussion again now, in the context of what we have said about the Newtonian paradigm. In particular, we will look more closely at the ways in which mathematical mappings, or functional relations, appear as images of physical reality in Newtonian modelling relations described by Fig. 1. By seeing explicitly what attributes of material reality in natural systems are actually encoded by mappings via the Newtonian paradigm, we will be in a stronger position to talk about realizations of mappings in the (M, R)-systems. But as we shall see, there are several surprises in store.

To see what is involved in the dual activities of encoding events into mappings, and realizing mappings as events, we will consider a few typical examples.

Our first example will be one that I have analyzed in detail elsewhere, a rather simple and degenerate (i.e., nondynamical) thermodynamic situation but one that illustrates clearly some of the basic issues. This is the van der Waals equation, a typical thermodynamic equation of state describing the equilibrium points of a class of nonideal gases. As originally formulated, this equation of state could be written in the form

$$(p + a/v^2)(v - b) = rT. \tag{3}$$

Here p, v, T are interpreted as the thermodynamic variables of state (pressure, volume, and temperature, respectively), and a, b, and r are *parameters*.

We shall not pause to discuss here the rather lengthy arguments required actually to allow us to consider Eq. (3) as an encoding of material reality; this requires, among other things, characterizing the manner in which events are converted to numbers through specific transducers (meters). Rather, we shall consider Eq. (3) as (1) a mathematical expression and (2) as an encoding, a symbolization, or a description of some material situation.

In purely mathematical terms, Eq. (3) is a functional relation involving six arguments:

$$\Phi(p, v, T, a, b, r) = 0. \tag{4}$$

Mathematically, all the arguments of this relation are equivalent; they are simply arguments. But as *encodings*, there is the most profound distinction in the *interpretation* of these arguments. This distinction is *lost* if we only look at the encoded mathematical version Eq. (3); it has been abstracted away. It is precisely what is being lost in the encoding process that we shall examine here.

Intuitively, the arguments in Eq. (3) fall into three classes:

(1) The *parameters* (a, b, r). These have to do with the particular kind (*species*) of gas under consideration (e.g., O_2, CO_2, NO, air ...). At this point, it is immaterial whether we say that the parameters (a, b, r) determine or are determined by this species. For this reason we shall call these parameters the *genome* of Eq. (3).

(2) A pair of the remaining arguments [say, (p, T)], whose values are determined by the character of the *environment* with which our gas is interacting. Normally, these variables are interpreted as those under the experimenter's *control*. The point is that the specific values or numbers assumed by these quantities are determined by processes that *do not obey* Eq. (3). We will therefore call these arguments (p, T) *environments* of Eq. (3).

(3) The remaining arguments of Eq. (3); in this case the volume v. At equilibrium, the value of v is completely determined by Eq. (3) when the genome (a, b, r) and the environment (p, T) are specified. For this reason we shall call this value the *phenotype* of Eq. (3).

Thus by looking informally at the nature of the encoding Eq. (3), we see the greatest possible distinctions among its arguments, distinctions that are entirely missing from the mathematical structure of Eq. (3) itself.

Let us see if we can get them back. We will rewrite Eq. (4) in two steps. First, instead of regarding Eq. (4) as a single function of six arguments, let us rather regard it as a three-parameter family of functions of three arguments:

$$\{\Phi_{abr}(p, v, T) = 0\}. \tag{5}$$

Here we are using the parameters we identified as *genomic* as *coordinates in a function space*, and not as arguments. This introduces a purely mathematical distinction into our encoding, destroying the misleading symmetry among the arguments of the van der Waals equation with which we started. Of course, *which* arguments we use in this fashion must be made part of the initial encoding.

Next, we shall rewrite the three-parameter family of (5) of *relations* as a three-parameter family of *mappings*, from environment to phenotypes:

$$\Phi_{abr}: (p, T) \to (v). \tag{6}$$

In this form, the distinction between *environment* and *phenotype* is manifested mathematically, as an integral part of the encoding itself. But, of course, *which* arguments are chosen as environment and which are chosen as phenotype must also be an integral part of the initial encoding process itself.

We have already noted that the van der Waals equation is a degenerate situation, pertaining to equilibrium values (i.e., limiting values for long times). We can inch toward dynamical encodings if we make one further reinterpretation of (6). Namely, we recall that the situation described by the van der Waals equation is: *whatever the initial volume* v_0, once genome and environment are specified, the ultimate phenotype is that value v given by (3). Thus we can finally rewrite (6) as a three-parameter family of *operators*, mapping "initial phenotypes" v_0 into phenotypes v:

$$v_0 \xrightarrow{\Phi_{abr}(p,T)} v. \tag{7}$$

If we do this carefully enough, we can actually use the map Φ_{abr} to generate a vector field on the space of phenotypes (i.e., a true dynamics). However, this dynamics is extremely degenerate in the present situation, a consequence of the fact that thermodynamics from which an equation of state like the van der Waals equation is taken is dynamically as singular as it can get (e.g., the manifold of equilibria, or critical points, where nothing can happen, is of the highest possible dimension). This extreme nongenericity is in fact typical of the situations with which conventional physics deals and is one of the reasons (we shall see others) that biology (among many other things) falls outside it and that it is most ironic for the physicist to fancy himself as a purveyor of "universal laws."

By these rewritings, we have embodied explicitly, in the mathematics into which our physics is encoded, at least some of the basic distinctions that were originally lost. In the process, we have gone from a single manifold with a relation on it to something like a fiber space, with genomes as base space and state spaces, or operators on phenotypes, as fiber. We shall not pause here to discuss the rather far-reaching ramifications of this picture, particularly those associated with the *stability* of the parameterized families Φ_{abr}, with evolution and development. We shall, however, touch on these matters, in another guise, in subsequent sections.

For the moment, we merely wish to mention that the forms (5)–(7) allow us to define a variety of *partial* maps. For instance, if we keep genome fixed, we get a mathematical relation between phenotype and environment. If we keep environment fixed, we get a mathematical relation between genome and phenotype. Each of these is mathematically *also a mapping*. But the *interpretations* or realizations of these partial mappings are completely different from one another, even in this utterly simple context. We begin to glimpse, then, some of the subtleties involved in the problem of realizing an abstract mathematical mapping. To do so, we need to know the whole encoding from which it comes.

This is, in fact, only the tip of one of the icebergs implicit in the Newtonian encoding. We shall look at this particular iceberg in more detail, from a new angle, in the next section. We shall then turn our attention to another, even bigger one.

IX. CAUSALITY

Since the distinctions that are our major interest become obliterated as a consequence of the abstractions inherent in the Newtonian encoding, we need another language to make them manifest. The only language that I have found appropriate for this purpose came from a most unexpected quarter, from the old Aristotelian doctrine of the categories of causation. It is a language that everyone believes has been made utterly obsolete by the advent of the Newtonian paradigm. However, we have already implicitly used this language heavily; as we shall see, it in fact permeates the discussion of the preceding section. Let us pause then to review briefly some of its essential features.

To Aristotle, all science is animated by a single question: Why? Science must answer this question "Why?"; it must say, "Because." In so doing, depending on the context, science becomes the vehicle for both explanation and prediction. Aristotle's basic contribution was to recognize that there are *different* and *inequivalent*, but equally valid, ways of saying "because." If we single out some event or thing in the external world and ask why it is what it is, then what we have singled out is the *effect* of its *causes*, and these causes are embodied in the different ways that Aristotle distinguished, in which we can answer the question we have asked.

Aristotle's explicit discussion of the categories of causation is, by modern standards, superficial and incomplete and, oddly, is couched primarily in terms of material artifacts. Following this discussion, let us suppose we are interested in "understanding" something like a house. The four Aristotelian categories of causation for the house are then:

(1) *Material cause*: The house is what it is *because* of the wood, bricks, glass, metal, and so on, of which it is composed. These constitute the material cause of the house and comprise one way of saying why the house is what it is.

(2) *Formal cause*: The house is what it is *because* of the blueprint or plan that it realizes. This blueprint or plan constitutes the formal cause of the house, and provides another way of saying why the house is what it is; a way different from, and inequivalent to, but equally valid as, the material cause.

(3) *Efficient cause*: The house is what it is *because* of the labor of its builders, who manipulated the constituent materials in accord with the blueprint or plan. We now have a third way of saying why the house is what it is; again different from, inequivalent to, but equally valid as the other ways.

(4) *Final cause*: The house is what it is *because* someone required shelter. This way of saying why was for Aristotle the most important because it involved *telos*; volition, goal, end. The study of the final causes of things is accordingly called *teleology*. Largely as a result of the Newtonian paradigm, the whole concept of teleology has become anathema; the corresponding adjective is as close to a defamation as there is in science.

What Aristotle was essentially doing in his discussion of the categories of causation was giving names to, and thereby distinguishing between, certain kinds of *relations between events*. As we noted in Section V, this is precisely the province of causality, one of the twin pillars supporting our belief in natural law. Nevertheless, it is widely believed that causality is not a scientific concept; that *true* science only began when Aristotelian ideas about causality were discarded and replaced by the Newtonian paradigm. Indeed, Bertrand Russell wrote an influential article pointing out precisely that the word *cause* never appears, as a technical term, in any "advanced science" such as "gravitational astronomy." For Russell, such "advanced sciences" consist *entirely* of mathematical relations (of the kind we have seen in the preceding section), which are *deterministic* without being *causal*. What Russell had done, in effect, was to forget about the whole left-hand side of the diagram in Fig. 1, including the encoding and decoding arrows. He tacitly took it as axiomatic that the Newtonian encoding was the only one; thus, we need never look again at the events themselves, and can content ourselves with looking exclusively at the mathematical images provided by that encoding. And of course it is perfectly true that, in this *formal* world of mathematical images, the word *cause* never appears; it has been lost in the encoding process, along with many other important things.

Nevertheless, we can restore the Aristotelian ideas by, as we did before, superimposing upon the Newtonian paradigm an additional informal layer of interpretation to compensate for the missing or unencoded properties we need. As we have seen, the essence of the Newtonian paradigm involves a *genome-parameterized family of environmentally determined operators, acting on a space of phenotypes*. If we now identify the space of phenotypes with the space of "internal states," or mechanical phases, it is not hard to show that the

Newtonian paradigm can be expressed in conventional mathematical language as a system of equations of motion for the phenotypes or states:

$$\frac{d\mathbf{z}}{dt} = \psi_{\mathbf{g}}(\mathbf{z}, \boldsymbol{\beta}(t)) \tag{8}$$

Here **z** is a state vector or phenotype vector; **g** is a genome vector, and $\boldsymbol{\beta} = \boldsymbol{\beta}(t)$ is a vector of environments, variously called *inputs* or *forcings* or *controls.* Relations such as this are precisely what Bertrand Russell *identified* with "advanced sciences."

Dynamical relations such as (8) can be converted into a mathematically equivalent form through a process of *integration*:

$$\mathbf{z}(t) = \int_{t_0}^{t} \psi_{\mathbf{g}}(\mathbf{z}, \boldsymbol{\beta}(\tau))\, d\tau + \varphi(\mathbf{z}(t_0)) \tag{9}$$

Although (8) and (9) are *mathematically* equivalent, they are *epistemologically* very different. The relation (8) is a local relation, relating the values of observable quantities at any given instant. The relation (9), on the other hand, relates the values of observable quantities at *different* instants. However, we shall not pursue the implications of this fact here.

Now let us look at the relation (9), which is one way of expressing the Newtonian paradigm, in terms of the Aristotelian ideas regarding categories of causation. Specifically, if we regard $\mathbf{z}(t)$, the phenotype or state of a system at some instant of time, as the *effect*, then we have:

$$\text{Material cause} \equiv \text{initial state } \mathbf{z}(t_0);$$

$$\text{Formal cause} \equiv \text{genome } \mathbf{g}.$$

$$\text{Efficient cause} \equiv \text{operator} \int_{t_0}^{t} \psi_{\mathbf{g}}(\ldots, \boldsymbol{\beta}(\tau)\, d\tau$$

We note explicitly that there is no *final cause* visible in this picture. In retrospect, it is this fact more than any other that has led to the profound belief that finality is incompatible with science. Indeed, any attempt to impose a category of final causation onto the Newtonian encoding destroys it completely. Insofar as finality involves the effect of future inputs, or future state, upon present change of state, the idea of *anticipation* has been expunged from serious science without further thought. The only exceptions I know involve some interesting discussions of, for example, advanced potentials as physically meaningful solutions of classical wave equations, and the interpretation of variational principles, which specify paths between prior and subsequent configurations, in anticipatory terms. But these are never taken seriously, for the reasons we have sketched already.

From what we have already said, innocent as it may appear, we can now reformulate the whole Newtonian paradigm in a way that manifests more clearly than any other how very special it in fact is. In brief, the essential fact is that *the Newtonian paradigm pertains to only those systems for which the categories of causation can be segregated into mathematically independent structures.* For instance, under the interpretation we have given, the very idea of a state space tacitly means that the category of material causation can be split off from the other causal categories as a completely independent unit; the same is true for all the other causal categories. Stated another way: Any observable quantity pertaining to a system, or to its environment, can be assigned exclusively to one or another of the causal categories, once assigned, it stays in that category, and can play no other causal role.

When cast in this light, we can see in a most vivid way that the whole Newtonian paradigm, which as noted earlier has persisted essentially unchanged from its inception as the only and universal mode of system description, is not that at all. Indeed, the class of special systems that it describes should be given a corresponding special name; we shall call such systems *simple systems* or *mechanisms.* The Newtonian paradigm tacitly says that *every system is a mechanism*; from the discussion we have given earlier, it is manifest that this need not be so. It will be the task of the following section to consider the question of whether there are natural systems that are not mechanisms and if so, how they are to be described.

For the remainder of the present section, however, we shall stay within the confines of the Newtonian paradigm (i.e., within the class of simple systems) and consider briefly a few of the implications of the fact that the categories of causation, in that paradigm, while segregated into independent mathematical structures, are nevertheless *inequivalent.* As far as I know, these implications have never been addressed, or even explicitly recognized, largely because of the intrinsic and historical peculiarities of the Newtonian paradigm itself.

A first conclusion that we can draw pertains to the idea of *simulation.* For instance, suppose we wish to simulate some dynamical system of the form (8), on an analog or a digital computer. This involves the construction of a diagram like that of Fig. 1 that encodes the sequence of state transitions in our original system (call it the *prototype*) into corresponding elements of the simulator. In terms of our preceding discussion, the sequence of state transitions in the prototype involves *material* causation in that system. However, in the simulator, state information about the prototype is always encoded as *data* or inputs to the simulator. Thus in the simulator the data are related to *efficient causation.* In fact, the whole idea of computation pertains essentially to the manipulation of efficient causation, with formal cause as *program* and material cause as "*hardware.*" Thus although a modelling relation can thereby be established between prototype and simulator in a *formal* way, the causal structures in the two are entirely different (and so too, incidentally, are the corresponding *inferential* structures).

It is generally believed that any dynamics can be simulated in this sense, i.e., that a modelling relation can always be established between material causation in some prototype and efficient causation in an appropriate simulator. Whether this is true or not I do not know. However, I do know that the reverse is false, that we cannot in general *realize* efficient causality in terms of material causality (hence their inequivalence). One well-known instance of this fact may be mentioned briefly here. Long ago (cf. Burks, 1966) John von Neumann gave a discussion of a putative "self-reproducing automaton." This discussion was based on the idea of a "universal constructor," which was in turn derived from Turing's earlier (1936) idea of a universal digital computer, or universal Turing machine. Basically, von Neumann's argument was that following a blueprint to construct something was just as much of an algorithmic process as following a program to compute something, and *therefore* that anything true of computation was necessarily equally true of construction. But of course a constructor must manipulate material causation, while a computer manipulates efficient causation; it does not follow, and is in fact false, that a universal simulator implies anything about a universal constructor. Indeed, for the same reason, one must be very careful in extrapolating from the properties of formal systems such as neural networks or automata back to the material properties of such processes as biological development or cellular control; for these too are exercises in the realization of efficient causation by material causation in some prototype. It was also for this reason that my first attempt to realize dynamically the (M, R)-systems (cf. Section V) did not work.

The same inequivalence between causal categories is manifested in the genotype–phenotype dualism, which animates the entire theory of evolution in biology and which in its essential features underlies any equation of state generated by a Newtonian encoding of a material system. That indeed is why we chose the terminology (genome, environment, phenotype) we used in classifying the arguments of such equations, and which as we have seen eventually manifests itself directly in causal terms. For instance, the D'Arcy Thompson "Theory of Transformations" (Thompson, 1917) asserts fundamentally that all closely related phenotypes are also similar; here "*closely related*" pertains to genotypes and *similar* pertains to phenotypes (or better, to environment–phenotype relations). Here the relation is between *formal cause* (the genomes) and efficient cause. The inequivalence of these two categories of causation means precisely, in modern mathematical language, that there exist bifurcating genomes, in any neighborhood of which there will be dissimilar phenotypes. This simple fact goes a long way to help us to understand the basis for what evolutionists call *macroevolution*, but that again is another story.

The inequivalence of causal categories manifests itself in many other ways. It is intimately involved, for instance, in the inability to infer anything much about, for example, the material structure of an enzyme from a knowledge of its substrates and products, and of a knowledge even of its kinetic parameters in the substrate-product conversion. The latter involves again efficient causation; the structure is material causation. More generally, it manifests itself in what is usually called the *system identification problem*, which can be translated precisely into an attempt to predict material cause from knowledge about efficient or formal cause. Going the other way, we cannot predict, for example, that a given material structure will play any particular kind of functional role (e.g., that a particular protein structure is an enzyme, let alone what its substrates are).

And even in pure mathematics itself, results like Gödel's incompleteness theorems are manifestations of this same inequivalence, now expressed in completely formal terms.

Thus we can now appreciate better some of the real difficulties involved in attempting to realize physically a relational structure such as an (M, R)-system. Indeed, if we look again at the (M, R)-system (1) in the light of the discussion of the past few sections, we can see just how complicated a little structure it is, in terms of its causal correlates. More precisely, the sets and mappings that comprise it, though all *mathematically* of a common character, have completely different causal interpretations; sometimes the same thing (e.g., the metabolic map f) is involved in several of these categories simultaneously. Indeed, it may well be (although I cannot prove it) that even such an elementary relational structure as this (M, R)-system in fact cannot be realized within the confines of the Newtonian paradigm at all.

We shall now turn to a consideration of the following question: If indeed the Newtonian paradigm is so special, what alternative is there? If the Newtonian paradigm describes only the limited class of systems that we have called mechanisms, or simple systems, how are we to describe those systems (if any) that fall outside that class? This will be the subject of the next section.

X. COMPLEX SYSTEMS

We now turn our attention to the question of what a mathematical image of a complex system should be like. That is, we suppose that there is a complex system sitting on the left-hand side of the diagram of Fig. 1 and ask what kind of mathematical object can go into the right-hand side so that the diagram will commute.

Before considering some explicit possibilities, we can already draw some general conclusions regarding these new mathematical images. Some of these are:

(1) There can be no such thing as a "state space" in such an image, which can be fixed once and for all. More generally, the causal categories (which become much more subtle in this context) cannot be segregated into disjoint classes; at least some elements of our image play several causal roles simultaneously. Moreover, these causal roles can shift in the course of time as a consequence of system dynamics.

(2) A complex system will have a multitude of partial images of the Newtonian type, which can in some sense "approximate" to the behavior of the system. But this approximation of complexity by simplicity is only local and temporary. This means that, as the complex system develops in time, any such simple approximation ceases to describe the system in the sense of Fig. 1; the *discrepancy* between what the complex system is actually doing (arrow 1 in the diagram) and the behavior of the simple approximation (arrows 2 + 3 + 4) grows in time. When the discrepancy becomes intolerable, we must replace our initial simple approximation by another. The discrepancy between the behavior of a complex system and any such simple approximation is, depending on the context, called *error* or *emergence.*

(3) Even though a complex system has a multitude of partial simple descriptions, we cannot construct from them a single "largest" description that is also simple. In this sense, the reductionistic paradigm fails for complex systems.

Just from these few properties, which follow essentially only from the non-Newtonian character of complexity, we can see that this kind of world of mathematical images must have very different properties from the one we are used to.

There are now two questions to answer: (1) Is there a world of mathematical structures with these characteristics, that can be put into a modelling relation with a complex system, and (2) are there complex systems in nature, which realize such mathematical structures? We shall turn to the first of these questions now, leaving the second for our final comments.

To motivate this discussion, let us return to a consideration of conventional Newtonian images of the form (8). More specifically, let us look at a dynamical system

$$dx_i/dt = f_i(x_1, \ldots, x_n) \tag{10}$$

leaving out of consideration for the moment the genomic and environmental aspects. Originally motivated by an attempt to establish some relation between dynamical and informational ways of treating such a system, and following an earlier treatment of Higgins (1967), I considered the (observable) quantities

$$u_{ij}(x_1, \ldots, x_n) = \partial/\partial x_i(dx_i/dt).$$

A great deal about the stability of (10) can be inferred from these quantities; in fact, most of the significance of the u_{ij} lies in their *signs*, and not so much in their specific values.

These quantities, as Higgins noted, have informational correlates. For instance, if u_{ij} is positive in a state, it means, for example, that an increase in x_j increases the rate of production of x_i (or that a decrease in x_j decreases the rate of production of x_i). It thus makes sense to call x_j an *activator* of x_i under these circumstances. Likewise, if u_{ij} is negative in a state, it is reasonable to call x_j an *inhibitor* of x_i. In fact, on this basis, we can convert the dynamical system (10) into an informational network, quite analogous to a neural net (cf. Rosen, 1979).

There are many situations in biochemical, morphogenetic, ecological, and neural theory in which the activation–inhibition language seems more natural than the dynamic one. Thus the obvious question was whether, given such an activation–inhibition network, we could in effect invert the preceding discussion and recover a set of rate equations (10). What we must do is straightforward; construct the differential forms

$$\omega_i = \sum_{i=1}^{n} u_{ij}\, dx_j. \tag{11}$$

If ω_i is exact, there is an observable f_i such that $df_i = \omega_i$. Put this observable equal to dx_i/dt and we are done. However, if these forms are *not exact*, we have an activation–inhibition network that cannot be realized by a set of rate equations.

For a differential form to be exact, if $n > 2$, is a most nongeneric situation. There are some standard necessary conditions for exactness, which may be written as

$$(\partial/\partial x_k)u_{ij} = (\partial/\partial x_j)u_{ik}.$$

Now these quantities

$$(\partial/\partial x_k)u_{ij} = u_{ijk}$$

also have an informational connotation. In brief, it is easy to see that, if u_{ijk} is positive in a state, it means that x_k enhances or potentiates the effect of x_j on x_i. Under these circumstances, we can call x_k an *agonist* of x_j. Likewise, if u_{ijk} is negative, x_k attenuates the effect of x_j on x_i, and we can call x_k an *antagonist* of x_j. We see then that the conditions for exactness of (11) become $u_{ijk} = u_{ikj}$ for all indices i, j, k; the activator–inhibitor relation and the agonist–antagonist relation are completely symmetrical. This too is a highly nongeneric situation. Thus we can conclude that the informational description is more general than that given by dynamical laws like (10).

In a nutshell, we can continue iterating the process whose first two steps we have described, constructing successive networks $u_{ij}, u_{ijk}, u_{ijkm}, \ldots$, each of which modulates the properties of its predecessors. If we start from a set of rate equations (10) all of these layers of networks are derivable from the rate equations; from any one we can reconstruct all the others in the obvious fashion. On the other hand, if any of the differential forms in these networks are inexact, the networks become independent of each other, and there is no set of rate equations from which all the layers follow.

Networks of this kind provide the first concrete examples of a class of mathematical images satisfying the requirements we have indicated earlier, for representing systems that are not simple. We cannot go into technical details in this short space, but we can set forth certain conclusions about these systems of layers of informational interactions:

(1) The class of all such images can be converted into a general mathematical structure called a category. In this category, the category of Newtonian images (i.e., of dynamical systems or state-determined systems, which are the images of simple systems or mechanisms sits as a very small subcategory, just as the rational numbers sit as a subset of measure zero in the set of real numbers. Moreover, just as in the case of the rational numbers, every object in the big category can be regarded as the limit of a sequence of elements in the small one. Thus we have the notion of the "approximation" of a complex system by a simple one; but as noted previously, this "approximation" is only local and temporary.

(2) The causal structure of the objects in the big category turns out to be much more complicated than is true in the subcategory of dynamical systems. The infinitely greater richness of the causal structures possible in complex systems provides one way to understand the growth of the deviation between what a complex system will do and what a simple approximation does. Moreover, this greater richness of causal structure makes the problem of interpretation or explanation of experimental observation very different from what we are used to.

(3) In complex systems, an ideal of *final* causation or *anticipation* can be introduced in a perfectly rigorous, nonmystical way. Briefly, a complex system may contain predictive models of itself and/or its environment, which it can utilize to modify its own present activities.

(4) Because complex systems ultimately depart from the behavior predicted on the basis of *any* simple approximation, their behavior appears to us to be surprising and *counterintuitive*.

There are many other conclusions to be drawn from the class of mathematical images that we have briefly described and their relation to the Newtonian ones that approximate to them. This relation, on the one hand, explains why we have been able to go as far as we have with the Newtonian paradigm and why, on the other hand, we can in many areas get no further. The relation between complex systems and their simple approximations may be likened to the situation faced by the early cartographers, who were attempting to map the surface of a sphere while armed only with pieces of planes. Here the sphere plays the role of a complex system, while a piece of (tangent) plane is like a simple approximation. As long as we only map local regions, the planar approximation suffices, but as we try to map larger and larger regions, the discrepancy between the map and the surface grows as well. As noted earlier, this discrepancy can be called either *error* (which can be located either in the sphere or in the planar map) or *emergence* (of a new property of the surface; namely, its curvature). Thus if we want to make accurate maps of large regions of the sphere, we have to keep shifting our tangent planes. The surface of the sphere is in some sense a limit of its planar approximations, but to specify it in this way requires a new global concept (the topology of the sphere; i.e., its curvature) that cannot be inferred from local planar maps alone.

XI. AN ALTERNATE APPROACH: "INFORMATION"

Because it is interesting and important in its own right and because it leads to an alternate mode of entry into the universe of complex systems that is of some independent interest, we shall turn in the present section to yet another analysis of the idea of "information." Ever since Shannon began to talk about "information theory" (by which he meant a probabilistic analysis of the deleterious effects of propagating signals through "channels"; cf. Shannon, 1949) this concept has been relentlessly analyzed and reanalyzed. The time and effort expended on these analyses must surely rank as one of the most unprofitable investments in modern scientific history; not only has there been no profit, but the currency itself has been debased to worthlessness. Yet in biology, for example, the terminology of information intrudes itself insistently at every level—code, signal, program, computation, recognition. It may be that these informational terms are simply not scientific at all, that they are an anthropomorphic stopgap, a *façon de parler* that merely reflects the immaturity of biology as a science, to be replaced at the earliest opportunity by the more rigorous terminology of force, energy, and potential that are the province of more mature sciences (i.e., physics) in which "information" is never mentioned. Or it may be that the informational terminology that seems to force itself upon us bespeaks something fundamental, something that is *missing* from physics as we now understand it.

In human terms, information is easy to define; it is anything that is or can be the answer to a question. Therefore we shall preface our more formal considerations with a brief discussion of the status of interrogatives, in logic and in science.

The amazing fact is that interrogation is never a part of formal logic, including mathematics. The symbol "?" is not a logical symbol, as, for instance, are "$\vee$," "$\wedge$," "$\exists$," or "$\forall$," nor is it a mathematical symbol. It belongs entirely to informal discourse, and as far as I know, the purely logical or formal character of interrogation has never been investigated. Thus if "information" is indeed connected in an intimate fashion with interrogation, it is not surprising that it has not been formally characterized in any real sense. There is simply no existing basis on which to do so.

I do not intend to go deeply here into the problem of extending formal logic (always including mathematics in this domain) so as to include interrogatories. What I want to suggest here is a relation between our informal notions of interrogation and the familiar logical operation "$\Rightarrow$"—the conditional, or the implication operation. Colloquially, this operation can be rendered in the form "If A, then B." My argument will involve two steps. First, I will argue that *every* interrogative can be put into a kind of conditional form:

If A, then B?

(where B can be an indefinite pronoun such as *who*, *what*, etc., as well as a definite proposition). Second, and most important, I will argue that every interrogative can be expressed in a more special conditional form, which can be described as follows. Suppose I know that some proposition of the form

If A, then B

is true. Suppose I now change or vary A, that is replace A by a new expression which I will call δA. The result will be an interrogative, which I can express as

If δA, then δB?

Roughly, I am treating the true proposition "If A, then B" as a reference, and I am asking what happens to this proposition if I replace the reference expression A by the new expression δA. I could of course do the same thing which B in the reference proposition, replace it by a new proposition δB and ask what happens to A. I assert that every interrogative can be expressed this way, in what I shall call a *variational form*.

The importance of these notions for us will lie in their relation to the external world, most particularly in their relation to the concept of *measurement* and to the notions of causality to which they become connected when a formal or·logical system is employed to represent what is happening in the external world (i.e., to describe some physical or biological system or situation) (cf. Section VI).

Before doing this, I want to motivate the two assertions made earlier regarding the expression of arbitrary interrogatives in a kind of conditional form. I will do this by considering a few typical examples and leaving the rest to the reader for the moment.

Suppose I consider the question

"Did it rain yesterday?"

First, I will write it in the form

"If (yesterday), then (rain)?"

which is the first kind of conditional form described earlier. To find the variational form, I presume I know that some proposition such as

"If (today), then (sunny)"

is true. The general variational form of this proposition is

"If δ(today), then δ(sunny)?"

In particular, then, if I put

$$\delta(\text{today}) = (\text{yesterday})$$

$$\delta(\text{sunny}) = (\text{rain})$$

I have indeed expressed my original question in the variational form. A little experimentation with interrogatives of various kinds taken from informal discourse (of great interest are questions of classification, including existence and universality) should serve to make manifest the generality of the relation between interrogation and the implicative forms described earlier. Of course this cannot be *proved* in any logical sense, since as noted earlier, interrogation sits outside logic.

It is clear that the notions of observation and experiment are closely related to the concept of interrogation. That is why the results of observation and experiment (i.e., data) are so generally regarded as being information. In a formal sense, simple observation can be regarded as a special case of experimentation; intuitively, an observer simply determines what *is*, while an experimenter systematically perturbs what is and then observes the effects of his perturbation. In the conditional form, then, an observer is asking a question that can generally be expressed as:

"If (initial conditions), then (meter reading)?"

In the variational form, this question may be formulated as follows: Assuming the proposition

"If (initial conditions = 0), then (meter readings = 0)"

is true (this establishes the reference and corresponds to calibrating the meters), our question becomes

"If δ(initial conditions = 0), then δ(meter readings = 0)?"

where simply

$$\delta(\text{initial conditions} = 0) = (\text{initial conditions})$$

and

$$\delta(\text{meter readings} = 0) = (\text{meter readings}).$$

The experimentalist essentially takes the results of observation as his reference and thus basically asks the question that in variational form is just

"If δ(initial conditions), then δ(meter readings)?"

The theoretical scientist, on the other hand, deals with a different class of question, namely, the questions that arise from assuming a δB (which may be B itself) and asking for the corresponding δA. This is a question that an experimentalist cannot approach directly, not even in principle. It is mainly the difference between the two kinds of questions that marks the difference between experiment and theory, as well as the difference between the explanatory and predictive roles of theory itself; clearly, if we give δA and ask for the consequent δB, we are predicting, whereas if we assume a δB and ask for the antecedent δA, we are explaining.

It should be noted that exactly the same duality arises in mathematics and logic themselves; i.e., in purely formal systems. Thus a mathematician can ask (*informally*): If (I make certain assumptions), then (what follows)? Or he can start with a conjecture and ask: If (Fermat's last theorem is true), then (what initial conditions must I assume to construct explicitly a proof)? The former is analogous to prediction, the latter to explanation.

When formal systems (i.e., logic and mathematics) are used to construct images of what is going on in the world, then interrogations and implications become associated with ideas of causality. Indeed, we have seen that the whole concept of natural law depends precisely on the idea that causal processes in natural systems can be made to correspond with implication in some appropriate descriptive inferential system.

But the concept of causality is itself a complicated one; this fact has been largely overlooked in modern scientific discourse, to its cost. That causality is complicated was already noted by Aristotle, when he pointed out that there were four distinct categories of causation, four ways of answering the question *why*. These categories he called *material cause*, *formal cause*, *efficient cause*, and *final cause*. We have already seen that these categories of causation are inequivalent; hence there are correspondingly *different kinds of information*, associated with different causal categories. These different kinds of information have been confused, mainly because we are in the habit of using the same mathematical language to describe all of them; it is from these inherent confusions that much of the ambiguity and murkiness of the concept of information ultimately arises. Indeed, we can repeat in the present context what we have already noted: The very fact that the same mathematical language does not (in fact, cannot) distinguish between essentially distinct categories of causation means that the mathematical language we have been using is in itself somehow fundamentally deficient and that it must be extended by means of supplementary structures to eliminate those deficiencies.

However, there is yet a deeper relation between information, interrogation, causality, and mathematics implicit in the preceding discussion. This relation has important consequences for the structure of mathematics itself. Let us introduce it by noting that there is an exact parallel between the Newtonian paradigm, with its partition of system description into states plus dynamical laws, and the structure of mathematical formalisms, with their corresponding partition into propositions and production rules, or rules of inference. We have already noted that the Newtonian paradigm cannot accommodate the Aristotelian category of final causation (cf. Section IX). It is for precisely the same reason that *logical or mathematical systems cannot accommodate interrogation*, on which we have based the idea of "information"; namely, *an interrogation or question always involves a telic aspect*. It is precisely this telic aspect that eludes capture in a simple system, whether that system be a real material system or a mathematical image of such a system. We can now see intuitively that any attempt to construct a logical or mathematical formalism big enough to accommodate interrogation will lead us again directly into the category of complex systems, this time by a purely formal route. Indeed, seen in this light, the famous Gödel theorems, to which we have already referred, are about the approximation of a complex formalism by simple ones; here a "complex" formalism is, roughly speaking, one big enough to encode *within itself* a question of the form, "if (p), then (provable)?"

It may not be out of place here to mention parenthetically that (1) the telic nature of interrogation and (2) the close relation between observation (experiment) and interrogation are at the root of the conundrums associated with many analyses of the measurement problem in quantum mechanics. Quantum mechanics is entirely a classical theory in its partition of the world into states plus dynamical laws. Thus from our point of view, it is entirely subject to the analysis we have provided. Its only novel feature (and it is, of course, a central one) is in its postulation of what constitutes a state and how such a state is related to what is actually measured.

To conclude this section, let us return briefly to the role of interrogation in the theory of complex natural systems. We recall again that information, for us, is the answer to a question and that questions can be put into what we called the variational form: If δA, then δB? The connecting bridge between these considerations and physics lies in the interpretation of a question in variational form and the general concept of *virtual displacement*. In this guise the abstract considerations we have developed have already played a central role in classical physics—in mechanics, field theory, and thermodynamics.

In mechanics, a virtual displacement is a small, imaginary change imposed on the *configuration* of a mechanical system, with the forces kept fixed. The animating question is: If such a virtual displacement is made, then what happens? The answer, in mechanics, is: If the mechanical system is at equilibrium, then the (virtual) work done by the impressed forces as a result of the virtual displacement must vanish. This principle of virtual work is a static principle (i.e., pertains only to equilibrium), but it can be extended from statics to dynamics, where it is known as *D'Alembert's principle* and leads directly to the equations of motion.

The reader will observe now that the informational networks of Section X are defined by functions that answer questions of the form posed by virtual displacements. For instance, a function such as

$$u_{ij} = \partial/\partial x_j(dx_i/dt)$$

answers a question such as

$$\text{If } \delta x_j, \text{ then } \delta(dx_i/dt)\,?$$

In fact, pursuing these considerations leads precisely back to the same informational nets as the ones we have seen before. But this time we do not have to detour through dynamical systems (i.e., simple systems) at all; we can proceed entirely through informational considerations, in the guise of questions about how a (virtual) displacement of an observable affects another. The interesting thing is that several quite independent approaches lead us back to precisely the same circle of ideas.

XII. CONCLUSION

We now turn to our final question: Are there any complex systems in nature? I would argue that biology is filled with them, that the most elementary relational considerations bring us instantly face to face with this fact. Like early man, who could see the rotation of the earth every evening just by watching the sky but could not understand what he was seeing, we have been unable to understand what every organism is telling us. It cannot be stressed strongly enough that the transition from simplicity to complexity is not merely a technical matter to be handled within the Newtonian paradigm; complexity is not just complication, to be described by another number (e.g., the dimension of a state space or the length of a program), but a whole new theoretical world, with a whole new physics associated with it.

If organisms are indeed complex in our sense and if contemporary physics deals exclusively with simple systems, it follows that we cannot in principle do biology within the confines of contemporary physics. This is simply a more precise statement of what we asserted earlier; that the relation of biology to our present physics is not that of particular to general. It is not biology but physics that deals with too limited, too restricted a class of systems. Far from biology being reduced to, and hence disappearing into, contemporary physics, as the reductionists believe, it is physics that will be transformed out of present recognition by being forced to confront, head on, the problems posed by complexity. The shambles that the concept of the "open system" has made of classical thermodynamics (where after 50 years or more there is still no real physics capable of properly coping with even the most elementary open system dynamics) is as nothing compared to the impact of complexity; and thermodynamics has long been regarded with complacency as the repository of the most universal truths of physics.

Let us express the situation outlined earlier in the form of a diagram. The Newtonian paradigm and its corollaries have told us that every science is a logical consequence of physics, i.e., of the science of mechanisms. The science of mechanisms is thus the root of a tree, with biology as one of the specialized branches, as follows:

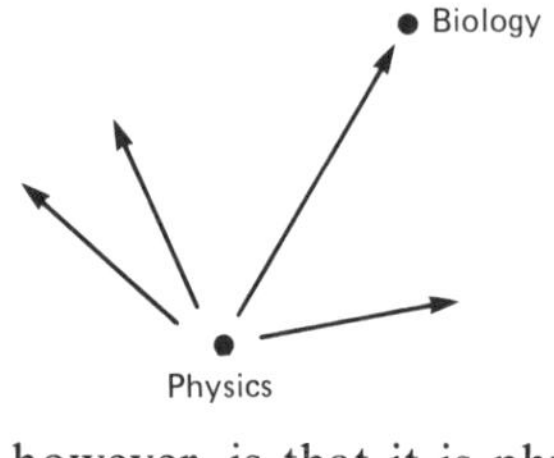

What we have argued, however, is that it is physics that is a specialized branch of a more general science of complex systems. Biology represents another branch, different from any science of mechanisms:

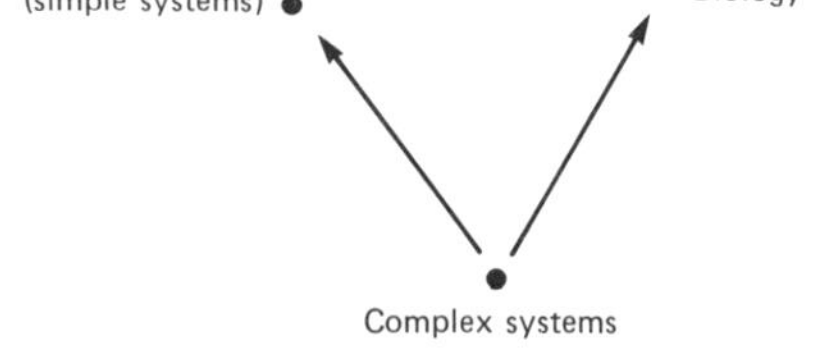

There are indeed relations between these two collateral branches; some of these we have tried to sketch earlier. But the relations are not reductionistic ones; they are more complicated, and more interesting, than that.

The power of the Newtonian paradigm, from this perspective, rests not in the fact that everything is a machine, but in the fact that complex systems, which by definition are not machines, can often act as if they were. We have much to learn about how this comes about, and even more about how it fails.

REFERENCES

Burks, A. (1966). "Theory of Self-Reproducing Automata." Univ. of Illinois Press, Urbana, Illinois.

Handler, Philip, Ed. (1970). "Biology and the Future of Man." Oxford Univ. Press, London.

Higgins, J. (1967). The theory of oscillating reactions. *J. Ind. & Eng. Chem.* **59**, 18–62.

Monod, Jacques (1971). "Chance and Necessity." English Translation. Knopf, New York.

Rashevsky, N. (1954). Topology and life. *Bull. Math. Biophys.* **16**, 317–348.

Rosen, R. (1964). Abstract biological systems as sequential machines. *Bull. Math. Biophysics* **26**, 103–111.

Rosen, R. (1979). Some comments on activation and inhibition. *Bull. Math. Biology* **41**, 427–445.

Shannon, C. (1949). "The Mathematical Theory of Communication." Univ. of Illinois Press, Urbana, Illinois.

Thompson, D'Arcy. (1917). "On Growth and Form." Cambridge Univ. Press, London and New York.

Turing, A. (1936). On compatible numbers. *Proc. London Math. Soc., Ser. 2* **42**, 230–265.

Index

G

H

I

K

L

M

N

O

P